Energy Systems in Electrical Engineering

Series Editor

Muhammad H. Rashid, Florida Polytechnic University, Lakeland, USA

More information about this series at http://www.springer.com/series/13509

Shanky Saxena · Ritu Sharma · B. D. Pant

Design and Development of MEMS based Guided Beam Type Piezoelectric Energy Harvester

Springer

Shanky Saxena
School of Electronics and Electrical Engineering
Lovely Professional University
Phagwara, Punjab, India

Ritu Sharma
Department of Electronics and Communication Engineering
Malaviya National Institute of Technology
Jaipur, Rajasthan, India

B. D. Pant
Council of Scientific and Industrial Research
Central Electronics Engineering Research Institute
Pilani, Rajasthan, India

ISSN 2199-8582 ISSN 2199-8590 (electronic)
Energy Systems in Electrical Engineering
ISBN 978-981-16-0605-2 ISBN 978-981-16-0606-9 (eBook)
https://doi.org/10.1007/978-981-16-0606-9

This Springer imprint is published by the registered company Springer Nature Singapore Pte Ltd.
The registered company address is: 152 Beach Road, #21-01/04 Gateway East, Singapore 189721, Singapore

This book is dedicated to my parents Mr. Anil Kumar Saxena and Mrs. Raagni Saxena for their continuous support and inspiration to pursue my interests in science. This book is also dedicated to my late grandparents Mr. P. L. Saxena, Mrs. Satyawati Saxena, Mr. M. C. Saxena and Mrs. Sharda Saxena for their love and encouragement in all my endeavors.

Shanky Saxena

Foreword

In the current race of miniaturization of microelectronic circuits and systems, the power required by these systems particularly by the Wireless Sensor Networks (WSNs) has reduced to a very low level. WSNs consist of many Mobile sensor nodes (MSNs) which are driven by miniaturized lithium-ion batteries that are exhaustible and replacement is not feasible at remote locations. Energy harvesting provides an unending power source that is adequate to power these sensor nodes specifically for remote area applications. Energy harvester can convert freely available ambient energy into electrical energy to provide an endless source of power for MSNs. In Microelectromechanical system (MEMS) based energy harvesters, electrical and mechanical components are integrated into a single device, making them tiny and compact which is suitable for remote area applications. MEMS-based piezoelectric vibration energy harvesters are a preferred choice because of their simplicity in design, operation, and compatibility with fabrication technology. The realization of cantilever-based piezoelectric energy harvesters has always been challenging because great care is required to develop and characterize each unit process leading to the technology development for the fabrication of a specific device.

This monograph is focused on MEMS-based piezoelectric vibration energy harvesters. State of the art technologies have been reviewed to understand the available technologies suitable for vibration energy harvesting. Analytical equations governing the device key parameters have been introduced in the initial portion of the monograph. Simulation steps have been presented using snapshots and results have been depicted by graphs making it easy to comprehend. In the later section of the monograph, a detailed fabrication process on the development of micromachining processes based on both wet and dry etching techniques has been discussed. Elaborate structural, material, morphological, topological, dynamic, and electrical characterization of the device has been presented. In the last section, a real-time practical application is selected, and a piezoelectric energy harvester is designed which is suitable to operate in the desired environment.

My compliments to the authors for putting together this valuable contribution in the form of a monograph which would be of interest and use to graduate students and researchers alike.

January 2021

Muhammad Mustafa Hussain, Ph.D.
Fellow, IEEE, APS and IoP
Department of Electrical Engineering
and Computer Sciences
University of California
Berkeley, CA, USA

Preface

The internet of things (IoT) is a network embedded with sensors, software and other technologies that allow to transmit and receive data. Increased use of sensors in IoT imposes a challenge to power these sensors operating at remote locations. The use of Lithium-ion batteries is not sustainable as they are not environment friendly and requires timely replacement. Energy harvesting is a technology that converts freely available ambient energy into electrical energy. Vibration energy harvesting is preferred because vibrations of different amplitude and frequency are freely available in the environment. Piezoelectric type vibration energy harvesters are prominently used due to their simple operation and compatibility with MEMS fabrication technology. Cantilever structure fixed from one end and free from the other are widely used to realize a Piezoelectric vibration energy harvester.

In this monograph, a fixed-fixed or guided beam Piezoelectric Vibration Energy Harvester has been designed, fabricated and tested for low-frequency operation. Device design has been done ensuring low-frequency operation, stable and reliable device operation and higher potential generation. Single beam, Cantilever array and guided beam structures have been designed and investigated using FEM. Fabrication technology is one of the most significant steps in the realization of a device. Guided two-beam and four-beam piezoelectric energy harvester devices have been fabricated. In both of these devices, a combination of wet and dry bulk micromachining techniques has been utilized. Aqueous TMAH etching for wet bulk micromachining, RIE (reactive ion etching) and DRIE (deep reactive ion etching) for dry bulk micromachining processes have been used.

Elaborate characterization of fabricated two-beam and four-beam devices has been done. Structural characterization has been done to determine the device dimensions and quality of the deposited materials. The resonance frequency of the device is experimentally worked out using a Laser Doppler Vibrometer (LDV) and is measured to be 466 Hz for the guided two-beam device and 515 Hz for the guided four-beam device. Finally, vibration bench testing is done for the measurement of electric potential and sensitivities of the two-beam and four-beam devices. Both the devices operate in the frequency range from 10 to 1000 Hz which is the desired frequency range for harvesting ambient vibrations.

In the last section, design and FEM analysis for a piezoelectric energy harvester have been done for a real-time practical application.

This monograph is aimed at graduate students and researchers working in the field of energy harvesting. Analytical equations governing the energy harvester operation have been discussed along with the stepwise FEM analysis which will be helpful for the students working in this field. Further, the mask layout design, fabrication and characterization techniques used are described in depth in the monograph using snapshots during each process which will be useful for researchers and engineers working in this field.

Ajmer, India Shanky Saxena
Jaipur, India Ritu Sharma
Pilani, India B. D. Pant

Acknowledgements

The authors would like to thank the scientific and technical staff from CSIR-Central Electronics Engineering Research Institute, Pilani, India, Materials Research Centre, Malaviya National Institute of Technology, Jaipur, India, Acoustics and Vibration standards laboratory, National Physical Laboratory, Delhi, India, Structural Solutions Pvt. Ltd., Hyderabad, India, Prof. V. Gupta and Dr. M. Tomar, Delhi University, Delhi, India for their support, without their expertise, this work would not have been possible. The authors acknowledge IEEE, Springer and Elsevier for granting permission to reuse materials copyrighted by these publishers in this monograph. The authors would like to express their deep sense of gratitude to their parents and teachers who have led the foundation and made them competent enough to write this monograph. We would like to thank Editor Prof. Muhammad H. Rashid, Associate editor, Ms. Kamiya Khattar, and staff of Springer, for their professional and efficient editorial work on this book. Finally, we would thank God for bestowing grace upon us, which helped us to keep going in the tough times and complete this work successfully.

Ajmer, India — Shanky Saxena
Jaipur, India — Ritu Sharma
Pilani, India — B. D. Pant

Contents

About the Authors

Shanky Saxena is an Assistant Professor in the School of Electronics and Electrical Engineering at Lovely Professional University, Punjab, India. He received a degree in electronics and communication engineering from the University of Rajasthan, India, in 2008, and a master's degree in communication and signal processing from Jaipur National University, Jaipur, India, in 2013. Dr. Saxena completed his Ph.D. from Malaviya National Institute of Technology, Jaipur, India, in 2019. He worked as a research scholar at the MEMS and Microsensors Department, CSIR-Central Electronics Engineering Research Institute, Pilani, India, from 2013 to 2018. He is a Senior Member of IEEE USA and has over 9 years of teaching and research experience with more than 25 publications in the MEMS field. His research interests include MEMS-based piezoelectric vibration energy harvesters and energy harvesting for bio-medical applications.

Ritu Sharma is an Associate Professor in the Department of Electronics and Communication Engineering, Malaviya National Institute of Technology (MNIT), Jaipur. She completed an M.Tech. in VLSI and a Ph.D. degree in photonics and optics from the Department of Electronics and Communication Engineering, MNIT, in 2004 and 2011, respectively. Dr. Sharma is working on important projects in the field of energy harvesters, optoelectronics and biosensors. Her research interests include fabrication and characterization of piezoelectric energy harvesters, nanophotonic devices based on low-cost materials, nanomaterials for nano-optoelectronic and nano-piezoelectric devices, MEMS, and nanoelectronic devices.

B. D. Pant worked as a Professor with AcSIR, and also an Emeritus Scientist with the Smart Sensors Area of CSIR-CEERI Pilani, India. He received an M.E. degree in microelectronics from the Birla Institute of Technology and Science (BITS), Pilani, India, in 1994, and a Ph.D. degree in MEMS from Kurukshetra University, Kurukshetra, India, in 2010. Dr. Pant has worked on several important projects of global interest e.g. the development of high-power piezoelectric materials, semiconductor devices, VLSICs, and MEMS. He has also led several projects on MEMS and NEMS. His interests are focused on developing MEMS technology and devices for medical, industrial, automotive, social security, and aerospace applications. His current interests are Bio-MEMS, microfluidics, micro, and nano-energy harvesters.

Abbreviations

AFM	Atomic force microscopy
DI	Deionized
DRIE	Deep reactive-ion etching
EMF	Electromotive force
FFT	Fast Fourier transform
LDV	Laser doppler vibrometer
MEMS	Microelectromechanical systems
MFC	Microbial fuel cell
MSN	Mobile sensor node
PB	Piezoelectric bimorph
PECVD	Plasma-enhanced chemical vapor deposition
PVDF	Polyvinylidene fluoride
P-VEH	Piezoelectric vibration energy harvester
PZT	Lead zirconate titanate
SEM	Scanning electron microscope
TMAH	Tetramethylammonium hydroxide
TPMS	Tire pressure monitoring system
WSN	Wireless sensor network
XRD	X-ray diffraction

Symbols

Al	Aluminum
AlN	Aluminum nitride
Au	Gold
b	Damping
C	Capacitance
E	Young's modulus
HCl	Hydrochloric
I	Moment of Inertia
k	Spring constant
l	Length of cantilever
m	Seismic mass
Pt	Platinum
Si	Silicon
SiO_2	Silicon oxide
t	Thickness of beam
W	Width of cantilever
x	Distance between plates
ZnO	Zinc oxide

Chapter 1
Introduction

Abstract This chapter introduces energy harvesting for wireless sensor networks. Mobile sensor nodes located at remote locations require power which can be provided through available renewable energy. State-of-the-art technologies used for energy harvesting in WSNs have been presented to understand the amount of power required by the sensor nodes and the energy source suitable to power them. Vibration energy harvesting and its transduction mechanisms, i.e., electromagnetic, capacitive and piezoelectric, used to scavenge vibration energy that can be realized in MEMS technology are discussed.

Keyword Wireless sensor network · Mobile sensor nodes · Piezoelectric · Vibration energy harvesters

1.1 Introduction

A wireless sensor network (WSN) consisting of a large number of randomly deployed mobile sensor nodes (MSNs) was initially introduced with a focus on intrusion detection for tactical applications (Agrawal 2017). Since then, WSNs have been used in diverse applications such as measurement of microclimates in agriculture farms, on road traffic monitoring, detection of human presence in workplaces and homes, and many more (Bhatti et al. 2016). WSNs consist of several different functional parts, i.e., a transducer, an analog-to-digital converter, a wireless transmitter/receiver and a battery to power the system. WSNs have remote area applications such as pressure monitors for car tires, wireless weather stations and implantable/mobile medical devices. These WSNs are powered by small lithium-ion batteries which are efficient but are not environmentally friendly and exhaustible (Knight et al. 2008). Replacement of these batteries typically in remote locations is practically not feasible. Moreover, these batteries make the overall system heavy through addition of its weight and volume, thereby lowering the overall efficiency. These batteries also adversely affect the environment when discarded without necessary precautions (Liu 2012a).

WSNs powered with natural resources such as thermal energy, wind energy, solar energy, vibration energy, etc. stand an effective solution for the aforesaid problem. Therefore, the main objective here is to design and develop an energy harvester device

S. Saxena et al., *Design and Development of MEMS based Guided Beam Type Piezoelectric Energy Harvester*, Energy Systems in Electrical Engineering,
https://doi.org/10.1007/978-981-16-0606-9_1

which can be embedded with MSNs that can be powered remotely for a longer period or ideally for infinitely long time. The above idea has evolved from the fact that the low-power nano- and microdevices have very low power requirements which can be effectively met using an energy harvester or micropower generator.

1.2 Energy Requirement in WSNs

Recent advancements in microelectronics have reduced the device power consumption to milliwatt or microwatt and even to the nanowatt levels through miniaturization (Ardila et al. 2013). The small size, low power consumption and remote operational capability of WSNs have gained a tremendous momentum of research in this area through the past decades. Table 1.1 summarizes some commercially available wireless MSNs along with their power consumption levels (Knight et al. 2008).

As already stated, a WSN consists of multiple functional units, i.e., transducer, A/D converter, battery, memory, embedded processor and a transreceiver as shown in Fig. 1.1. A transducer or a sensor converts different form of energy (thermal, mechanical, etc.) to electrical energy and holds the responsibility of measuring the physical parameters of the sensor locality (Agrawal 2017).

Table 1.1 Comparison of power consumption for commercially available MSNs (Knight et al. 2008)

Node name	Sleep mode (μA)	Transmit mode (mA)	Receive mode (mA)	Duty cycle (mA)	Operating voltage (V)	Batteries required	Battery life (days)
Fleck3	80	36.8	18.4	0.27	3.3	3 no.	440
XBee™	10	45	50	0.51	2.8	2 no.	230
MICAz™	–	–	–	0.70	2.7	2 no.	170

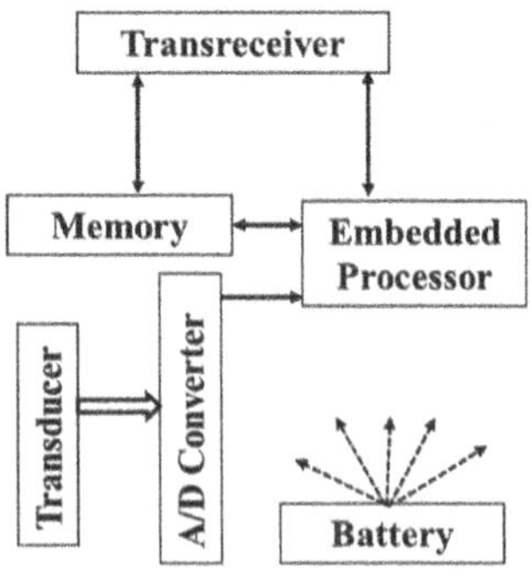

Fig. 1.1 WSN schematic depicting various functional units (Agrawal 2017)

Table 1.2 MICA2 motes SN energy consumption (Agrawal 2017)

Component	Current (mA)	Power
Nothing test	0.0075	23.25 nW
Radio off	0.004	12.4 nW
Radio idle	16	49.6 μW
Radio receiver	16	49.6 μW
Radio transmit	21	65.1 μW
Consumption alone	0.010	31.0 nW
Transmission cost	–	0.001 mJ/bit

The output from a sensor is in the form of analog current or voltage, which requires conversion in digital form and is done with the aid of analog-to-digital converter. This digital signal output from the A/D converter then gets processed by a dedicated embedded processor. The wireless transmitter/receiver that transmits or receives data wirelessly at any time of interest becomes the heart of the WSN. All these individual components/units get powered by small batteries embedded within the system. Commercial version of a SN is named as 'mote' (Agrawal 2017). Table 1.2 gives the power required for different component of SN.

1.3 Energy Harvesting for WSNs—State of the Art

"Energy harvesting is a technology that converts the excess energy available in the environment into usable energy for low power electronics" (de Queiroz 2013). Energy harvesting from environmental sources can be integrated with WSNs to enable perpetual operation and prolong the system lifetime. The major techniques that can be used conveniently in the WSNs mainly include light energy, heat energy, energy from vibrations, etc. (Bhatti et al. 2016). Based on the energy source, the state-of-the-art technologies used for energy harvesting in WSNs are being presented here.

Popular kinetic energy sources are mechanical stress, vibrations from manufacturing machines and sound waves. Vibration energy finds applicability in devices whose operational principle rely on electromagnetic, electrostatic and piezoelectric effects (Chye et al. 2010). Sazonov et al. (2009) in 2009 developed an energy harvester based on an electromagnetic approach which was employed for bridge health monitoring systems that can generate 12.5 mW from vehicular traffic vibrations. Rajamani and Vijayaraghavan (2010) in 2010 implemented a self-sustainable WSN that effectively monitors the traffic flow utilizing the vibration energy of the passing vehicles. On board, sensors can also be powered using vehicular vibrations. Piezoelectric harvesters utilizing vehicular vibrations were used by Talampus and Tolentino (2012) in 2012 and designed a self-powered vehicle tracker. Electrostatic harvesters were used by Löhndorf et al. (2007) in 2007 to evaluate their applicability in self-sustained tire pressure monitoring system (TPMS) ensuring safety of trailer trucks and heavy vehicles. Dondi et al. (2012) in 2012 developed a WSN comprising of different magnetometers and accelerometers which employs piezoelectric energy harvesters that get self-powered and also powers the nodes from trailer-induced vibrations.

Extraction of energy from electromagnetic radiations below infrared spectrum is feasible as per photoelectric effect taking visible/solar energy as a primary source. Solar cells or photovoltaic cells comprising of two layers of semiconductor (commonly Si) material is conventionally used for energy harvesting using photoelectric effect. Solar cells extending the battery life of MSNs in an automated and self-sustainable irrigation system were demonstrated by Gutiérrez et al. (2014) in 2014. Further, Brunelli et al. (2008) in 2008 enhanced the solar cell performance through a 1 mW energy harvesting tracking module at microlevels. Yerva et al. (2012) in 2012 demonstrated a sensor capable of harvesting indoor solar energy sufficient enough to transmit and receive minute-wise sensor data through a pre-defined collection route tree.

Radio frequency energy transmitted from Wi-Fi network, FM radio and cellular stations can be harvested. Cong et al. (2009) in 2009 designed and demonstrated a self-sustainable blood pressure monitor with dimensions in millimeter scale that gets powered from external RF sources. Similarly, another micrometer-scale cubic sensor got demonstrated by Sun et al. (2012) in 2012 with a rectenna capable of harvesting 1 μW of power to be used in implantable sensors.

Thermoelectric energy harvesters, based on temperature difference, got employed to power WSN self-powered node by Zhao et al. (2014) in 2014. Structural health monitors for water, oil and gas pipes have been demonstrated by Zhang et al. (2011) in 2011, and Martin et al. (2012) in 2012 used thermal sources in the environment such as steam and hot water. Rizzon et al. (2013) in 2013 utilized energy from heat dissipated by CPUs in data centers, in order to run dedicated WSNs meant for environment monitoring. Thermal energy harvesters can simultaneously act as harvesters and sensors as presented by Campbell et al. (2014) in 2004, and Lossec et al. (2013) in 2013 suggested that medical sensors can be powered using the warmth of animal or human bodies.

Energy harvesting may be accomplished from chemical or biological energy depending on the environment of interest. Microbial fuel cells (MFCs) powered small oceanographic sensors presented by Reimers et al. (2001) in 2001 generated a constant output power of 0.05 W/m^2. Further, Donovan et al. (2008) in 2008 demonstrated management system for power boosting the voltage of these fuel cells making them suitable for use in wireless temperature sensors ensuring extended lifetimes of at least a year in natural water. Gong et al. (2011) in 2011 demonstrated a footprint MFC of 0.25 m with an average output power of 44 mW/m^2 powering an acoustic modem that transmits measured temperature values over 50 days. Table 1.3 summarizes the different energy harvesting techniques based on the ambient source.

From the Table 1.3, it is clear that the kinetic energy from vibrations is highly suitable for harvesting energy for WSNs. Therefore, we further focus our investigation on vibration-based energy harvesters.

Table 1.3 Classification of energy harvesting techniques based on ambient source and their key characteristics suitable for WSN (Bhatti et al. 2016)

Energy source		Transduction mechanism	Application	Output power	References
Kinetic	Vibrations	Electromagnetic	Structural health monitoring	12 mW	Sazonov et al. (2009)
	Vibrations	Piezoelectric	Safety of vehicles	13.5 μW	Löhndorf et al. (2007)
	Vibrations	Piezoelectric	Safety of agriculture machinery	724 μW	Scorcioni et al. (2011)
	Air flow	Electromagnetic	Detection of forest fire	7.7 mW (@3.6 m/s)	Tan and Panda (2011)
	Air flow	Electromagnetic	Automatic HVAC	45 mW (@9 m/s)	Sardini and Serpelloni (2011)
	Water flow	Electromagnetic	Water pipe monitoring	18 mW	Morais et al. (2008)
	Human motion	Piezoelectric	Biomedical implants	1 mW	Almouahed et al. (2011)
	Animal motion	Electromagnetic	Herd localization	N/A	Dopico et al. (2012)
Radiant	Indoor light	Photovoltaic	Indoor applications	180 μW	Roundy et al. (2003)
	Sunlight	Photovoltaic	Smart irrigation	240 mW	Gutiérrez et al. (2014)
	Ambient RF	Rectenna	Outdoor sensing	60 μW	Sample et al. (2011)
Thermal		Thermoelectric	Environment monitoring	218 μW	Rizzon et al. (2013)
		Thermoelectric	Water metering	250 μW	Campbell et al. (2014)
Biochemical		Microbial fuel cell	Precision agriculture	310.24 μW	Pietrelli et al. (2014)

1.4 Vibration-Based Energy Harvesters

Vibrations are a reliable source of energy due to their free availability in the environment having different amplitude and frequency (Gherca and Olaru 2011). The vibration frequency range present in the environment suitable for vibration energy harvesting is given below in Table 1.4.

Most vibration energy harvesters (also called vibration-based microgenerators) are modeled as single degree-of-freedom second-order spring–mass systems (see Fig. 1.2) as first described by Williams and Yates (1996). The system consists of an

Table 1.4 Low-frequency vibrations

Application domain	Vibration frequency range (Hz)	References
Motor vibrations (car, engines)	30–300	Jha et al. (1976)
Human body vibrations	~400	Koenig et al. (2008)
Chopper vibrations	500–2000	Szefi et al. (2004)

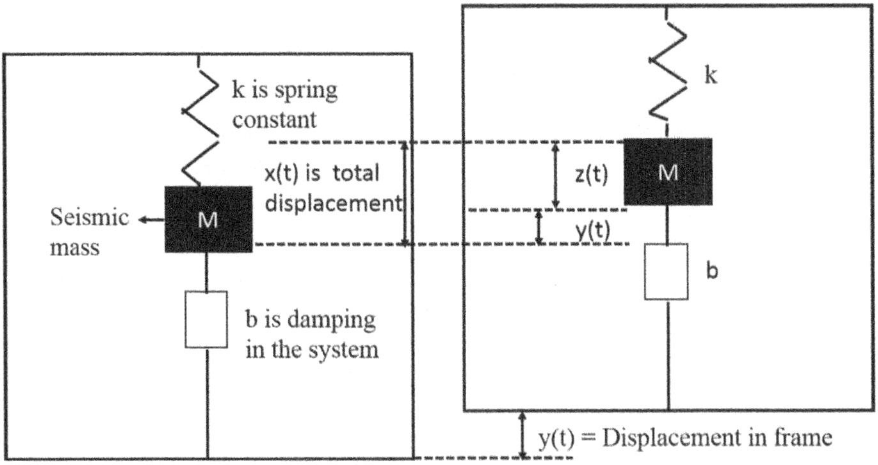

Fig. 1.2 Generic model of a vibration-based microgenerator (Williams and Yates 1996)

inertial frame that transmits the vibration to a suspended inertial mass, *m*, via the spring, *k*, and the damper, *b*, to produce a relative displacement or cause mechanical strain.

Figure 1.2 shows an inertial structure with simple spring-mass-damper system attached to a fixed frame. If generator housing vibrates with a displacement $y(t)$, which displaces the mass by $z(t)$ due to relative motion with respect to housing, then the differential motion is given by Eq. 1.1.

$$m\ddot{z}(t) + b\dot{z}(t) + kz(t) = -m\ddot{y}(t) \tag{1.1}$$

The force on the mass is equal to the force on the mass–spring–damper given by Eq. 1.2.

$$F = -m\ddot{y}(t) \tag{1.2}$$

The power transfer in the mass, $p(t)$, is the product of the force on the mass and its velocity given by Eq. 1.3.

$$p(t) = -m\ddot{y}(t)[\dot{y}(t) + \dot{z}(t)] \tag{1.3}$$

The motion of the mass relative to the fixed frame due to damping explains the basic principle of operation of the device. The relative displacement created with respect to the housing results from the vibration of the mass–spring system which, in turn, is due to the vibration of the fixed frame or housing due to the environmental vibrations. Resonance frequency when matches with the excitation frequency, then the mass stroke is magnified by the total damping ratio of the system and a considerable amount of energy is converted to useful electrical energy (Williams and Yates 1996).

Vibration energy harvesters are realized through MEMS technology because of the fact that in MEMS technology, electrical and mechanical components are integrated in the form of tiny devices or systems (Liu 2012b). This enables the integration of the mechanical moving parts such as cantilever beams with the electrical and electronic parts such as electronic circuits used for signal conditioning. These systems having sizes varying from micrometers to millimeters are fabricated by making use of standard batch processing techniques, have got inherently the ability to function as controllers, actuators and sensors in microscale by proper processing or amplification. MEMS devices or systems gain advantages from the excellent mechanical and electrical properties and stability of Si, whereas the ICs exploit only its electrical properties. MEMS, in general, comprises of mechanical microsensors, microactuators, microelectronics and microsensors all integrated into a single silicon chip (Bao 2005). SNs used in WSN are also MEMS-based sensors; therefore, integration of these SNs with a MEMS-based vibration energy harvester is highly suitable.

Three transduction mechanisms, i.e., electromagnetic, capacitive and piezoelectric, are used to scavenge vibration energy that can be realized in MEMS technology (Dallago et al. 2010). To couple the environmental vibrations to the transduction method, vibration energy harvesting requires a transducer and an energy converter. Such transducers must be designed carefully to maximize the energy converted with minimum loss.

The electromagnetic-type energy harvesters operate on the principle of Faraday's law of electromagnetic induction, i.e., induction of electro motive force (emf) in a coil due to change in magnetic flux linked with the coil (Soliman et al. 2006). A basic prototype of electromagnetic-type energy harvester is shown in Fig. 1.3. There are two fixed magnets, one each at the upper and lower sides. Under external vibration, the movable magnets get displaced from their mean position which results in the change in magnetic flux in the coil. According to Faraday's law of electromagnetic induction, emf is induced in the coil shown in Fig. 1.3. In the closed circuit, induced current will flow through the coil which is the harvester output current.

This type of design can also be implemented in MEMS by using a cantilever structure. A coil can be wound over a cantilever which is placed between magnets; during external vibration, cantilever will be displaced which will induce current in the coil. Thus, by using permanent magnets, resonating cantilever beam and a coil, achievement of electromagnetic induction becomes feasible (Yuan et al. 2010). One among the two, the magnet or coil will be kept fixed while the other would be

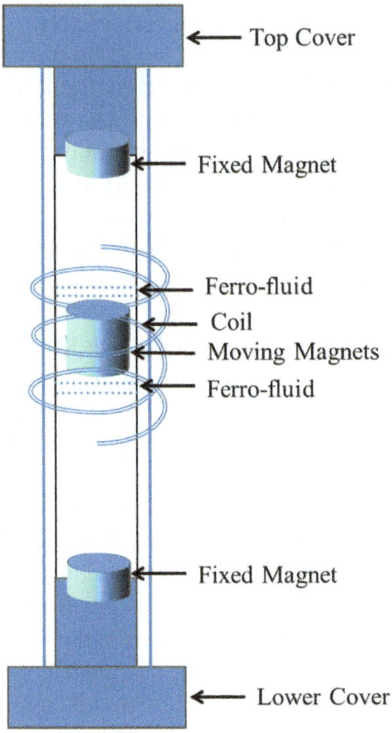

Fig. 1.3 Electromagnetic energy harvester

mounted on a beam. However, it is more preferable when the magnet is attached to the beam instead of the coil as they have the capability to act as inertial mass. The resonant frequencies of such systems have to be carefully designed to match the application environment specific characteristic frequency so that the amplitude of the environment vibration gets amplified by the quality factor linked with the resonant system (Li et al. 2013).

Electrostatic-type energy harvesters are based on the principle of constant charge separation. Increasing the electrical potential of charge by moving a fixed charge along an electric field is one of the methods of operation of an electrostatic transducer (Chamanian et al. 2012). In a parallel plate capacitor, under external vibration when the distance between the plates is reduced the capacitance is changed which results in the charge variation. Figure 1.4 shows a fringe type of arrangement of fixed electrodes and movable electrodes. Under external vibration, the plates move in plane x-direction which results in capacitance variation and charge generation. Electrostatic field developed will be independent of the plate separation and will be directly proportional to charge involved, which is fixed. The mechanical work is done, which would be against the attractive force existing between the parallel plates, and leads to an increase in separation between the plates and additional storage of electrical/potential energy. When a force acts on the plates and the plates are moved

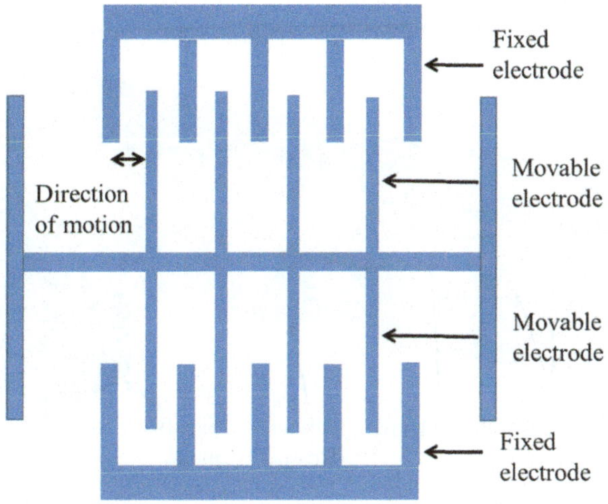

Fig. 1.4 Electrostatic energy harvester

relatively maintaining a fixed/constant separation, the work done will be against the existing fringing fields leading to an increment in electric field (Vullers et al. 2009).

The reduction in plate overlap leads to an enhancement in the stored electrical energy as field density increases at a faster rate than decrease in volume of the field as it is proportional to the square of the absolute magnitude of field strength. Some initial amount of energy becomes an essential requirement in all electrostatic transducers to generate the resultant higher electrical energy which is a demerit of using them in energy harvesting applications (O'Riordan et al. 2016).

Piezoelectric vibration energy harvesters (P-VEH) are based on the phenomenon of piezoelectric effect which was discovered by J and P Curie (Nicolson 1919). Piezoelectric polarization fields are generated when the mechanical strain in material results in production of an electric field across it, and the degree of polarization directly depends on the applied strain. The mechanical work extended on a strained material gets stored as elastic energy. Polarization fields developed in the material result in potential generation. Piezoelectric materials are widely being used in the form of piezoceramics (PZT), thin films (sputtered ZnO, AlN), polymeric materials (PVDF) and single crystals (quartz). As fabrication of magnets and coils in MEMS is practically difficult; therefore electromagnetic-type energy harvester is less suitable. Fabrication of capacitive type is suitable in MEMS but requires initial voltage for device operation. Piezoelectric type is best suitable for vibration energy harvesting due to its simple design, operation and because of ease of fabrication in MEMS technology. Piezoelectric energy harvesters are realized by cantilever structure fixed at one end and free at the other end having seismic mass as shown in Fig. 1.5. The base structure is of silicon having a beam and a seismic mass. Bottom and top electrodes' layer is deposited to collect charge generated. A piezoelectric layer is sandwiched between the electrodes which generates charge due to displacement in the cantilever

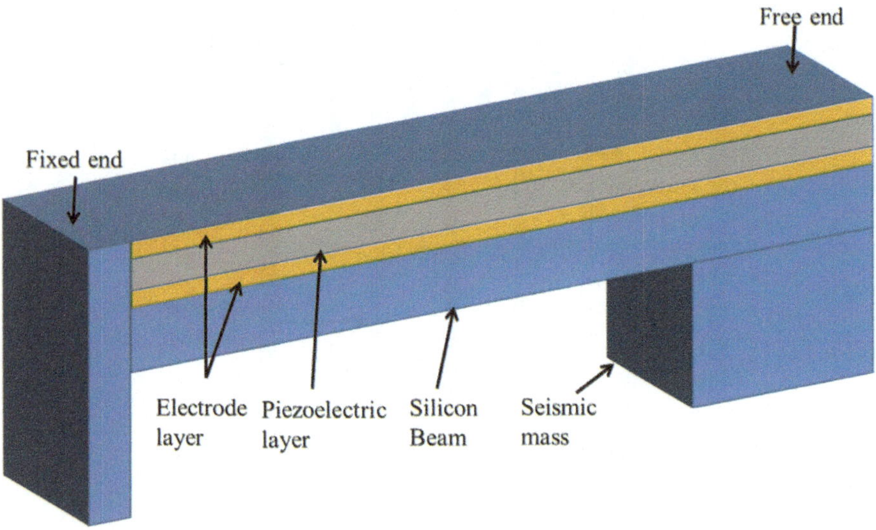

Fig. 1.5 Cantilever beam fixed from one end and seismic mass at other

under external vibration. One end of the cantilever is fixed, and the other is free as shown in Fig. 1.5.

In order to have a high sensitivity, the stiffness (spring constant k) of the cantilever has to be very low. However, lower stiffness of the cantilever makes it fragile. Therefore, for high-amplitude vibrations, a fixed-fixed beam structure which is fixed from both ends and having seismic mass at the center has been used for better stability and reliability (Marzencki et al. 2009).

Cantilever structures are modeled as spring–mass–damper system having a fixed resonance frequency. This resonance frequency depends on the physical parameters of cantilever such as length 'l', width 'W', thickness 't' and Young's modulus 'E' of the cantilever beam and weight of the seismic mass attached with the cantilever structure. The spring constant of the cantilever is therefore given by Eq. 1.4 (Zhu et al. 2009).

$$k = \frac{F}{x} = \frac{3EI}{l^3} = \frac{EWt^3}{4l^3}$$

The first criterion to generate maximum output power from the energy harvester is to tune the resonance frequency of the cantilever beam to the frequency of ambient vibration. The resonance frequency of the cantilever depends on spring constant (k) of the beam and seismic mass (m) at the center as given by Eq. 1.5 (Zhu et al. 2009).

$$f = \frac{1}{2\pi}\sqrt{\frac{k}{m}} \tag{1.5}$$

A cantilever-type P-VEH should operate in resonance with the ambient frequency for optimal output; otherwise the power harvested will drastically reduce and will not be able to drive sensor nodes (Karami et al. 2011). P-VEHs must be designed and operated in low-frequency region as ambient vibrations available in environment are in low-frequency range. Therefore, lower-frequency operation and resonance with ambient frequency vibration are the key parameters in successful operation of cantilever-type piezoelectric energy harvesters. The main advantage of P-VEHs is the compatibility of fabrication of cantilever structures in MEMS technology (Jamain et al. 2014). The cantilever structure shown in Fig. 1.5 can be fabricated on silicon substrate using bulk micromachining at the back side to realize proof mass and piezoelectric layer deposited on the front side with bottom and top electrodes. Ambient vibration displaces the cantilever from the free end which generates stress on the fixed end. The stress in the piezoelectric layer generates charge in the piezoelectric layer which results in the generation of electric potential. The potential generated is harvested from bottom and top electrodes. Reliability and stability can be ensured by making use of a guided beam design in which the cantilever beams are fixed at one end and supported with a seismic mass at the other end (Marzencki et al. 2009). The fixed nature of the beams leads to lower residual stress while vibrating as compared to single cantilever-based designs, enhancing the device stability and reliability. Guided beam structure gives lower displacement which reduces the possibility of collisions with the encapsulation layer, thereby preventing beam fractures improving the device reliability and stability.

1.5 Motivation

In the current scenario, there is a vast requirement for MEMS-based energy harvesters in WSNs. Some applications include structural health monitoring, tire pressure monitoring system, remote weather stations and many others. All these applications have a vibration frequency range from few Hz to KHz range as discussed earlier. Thus, to provide a clean and an unending energy source which can power these MSNs is the motivation of this work. Therefore, we aim to design and develop a MEMS-based piezoelectric-type vibration energy harvester.

1.6 Overview of Monograph

The remaining chapters of this monograph are organized as follows:

Chapter 2 presents the state-of-the-art technologies for piezoelectric vibration energy harvesters which give the current status of the cantilever-based piezoelectric-type energy harvesters, current design issues for the device fabricated in MEMS technology. Moreover, it helps in identifying the problems faced during practical energy harvesting applications for WSNs.

Chapter 3 presents the design and modeling of piezoelectric-type vibration energy harvester using FEM COMSOL Multiphysics®. Single beam, Cantilever Array and Guided beam structures are designed and results are compared. Analytical equations describing displacement, resonance frequency are presented. Effect of shape of seismic mass on potential generated by cantilever-type piezoelectric energy harvester is investigated.

Chapter 4 presents the simulation-based analysis of guided beam-type piezoelectric energy harvester. COMSOL Multiphysics® simulation tool is used to simulate guided two-beam- and four-beam-type piezoelectric energy harvester. Various datasets for the displacement, stress, potential generated and resonance frequency for the guided two-beam and four-beam structure at different input acceleration range are obtained, and results are compared. Dimensions of split electrodes have also been optimized which give reduced resonance frequency.

Chapter 5 presents fabrication of guided two-beam and four-beam P-VEH using MEMS fabrication technology. Bulk micromachining is done using tetra-methyl ammonium hydroxide (TMAH) CMOS compatible 25 wt% wet etching to realize pyramidal-shaped seismic mass, and deep reactive ion etching (DRIE) is done from front side to release two-beam and four-beam device. A piezoelectric layer of zinc oxide (ZnO) is sandwiched between top and bottom electrodes to collect charge generated under external vibrations. Finally, packaging of the fabricated device is presented.

Chapter 6 presents characterization of the guided two-beam and four-beam P-VEH. Structural characterization is done in which phase orientation of materials is done using X-ray diffraction (XRD), surface morphology is obtained using scanning electron microscope (SEM), and the topological view is obtained with atomic force microscopy (AFM). Dynamic characterization is performed using a laser Doppler vibrometer (LDV) to experimentally obtain the resonance frequency of the two-beam and four-beam device. Finally, vibration bench testing is done for measurement of electric potential and sensitivity for the two-beam and four-beam device. The variation of the voltage output with acceleration and frequency is also tested experimentally.

Chapter 7 gives the design and development of P-VEH for practical applications. Low frequency range is selected for design and FEM analysis of a fixed-free-type cantilever-based P-VEH. Effect of each design parameters such as length, width and thickness of the beam on the cantilever deflection, stress and electric potential generated is investigated. Parametric variation study is done where each parameter is increased individually and effect on the device performance is investigated. This study enables to optimize the device parameters and select the dimensions which are best suitable for the specific application.

References

Agrawal DP (2017) Applications of sensor networks. In: Embedded sensor systems. Springer, Singapore, pp 35–63

Almouahed S, Gouriou M, Hamitouche C, Stindel E, Roux C (2011) The use of piezoceramics as electrical energy harvesters within instrumented knee implant during walking. IEEE/ASME Trans Mechatron 16(5):799–807

Ardila G, Hinchet R, Montes L, Mouis M (2013) Mechanical energy harvesting with piezoelectric nanostructures: great expectations for autonomous systems. In: Future trends in microelectronics: frontiers and innovations, pp 230–243

Bao M (2005) Analysis and design principles of MEMS devices. Elsevier, Amsterdam

Bhatti NA, Alizai MH, Syed AA, Mottola L (2016) Energy harvesting and wireless transfer in sensor network applications: concepts and experiences. ACM Trans Sens Netw (TOSN) 12(3):24

Brunelli D, Benini L, Moser C, Thiele L (2008, March) An efficient solar energy harvester for wireless sensor nodes. In: 2008 design, automation and test in Europe, pp 104–109. IEEE

Campbell B, Ghena B, Dutta P (2014, November) Energy-harvesting thermoelectric sensing for unobtrusive water and appliance metering. In: Proceedings of the 2nd international workshop on energy neutral sensing systems, pp 7–12. ACM

Chamanian S, Zangabad RP, Zarbakhsh P, Bahrami M, Khodaei M (2012, February) Wideband capacitive energy harvester based on mechanical frequency-up conversion. In: 2012 IEEE sensors applications symposium proceedings, pp 1–4. IEEE

Chye WC, Dahari Z, Sidek O, Miskam MA (2010, October) Electromagnetic micro power generator—a comprehensive survey. In: 2010 IEEE Symposium on Industrial Electronics and Applications (ISIEA), pp 376–382. IEEE

Cong P, Chaimanonart N, Ko WH, Young DJ (2009) A wireless and batteryless 10-bit implantable blood pressure sensing microsystem with adaptive RF powering for real-time laboratory mice monitoring. IEEE J Solid-State Circuits 44(12):3631–3644

Dallago E, Marchesi M, Venchi G (2010) Analytical model of a vibrating electromagnetic harvester considering nonlinear effects. IEEE Trans Power Electron 25(8):1989–1997

de Queiroz ACM (2013) February. Electrostatic generators for vibrational energy harvesting. In: 2013 IEEE 4th Latin American Symposium on Circuits and Systems (LASCAS), pp 1–4. IEEE

Dondi D, Napoletano G, Bertacchini A, Larcher L, Pavan P (2012, October) A WSN system powered by vibrations to improve safety of machinery with trailer. In: SENSORS, 2012 IEEE, pp 1–4. IEEE

Donovan C, Dewan A, Heo D, Beyenal H (2008) Batteryless, wireless sensor powered by a sediment microbial fuel cell. Environ Sci Technol 42(22):8591–8596

Dopico NI, Gutiérrez Á, Zazo S (2012) Performance assessment of a kinetically-powered network for herd localization. Comput Electron Agric 87:74–84

Gherca R, Olaru R (2011) Harvesting vibration energy by electromagnetic induction. In: Annals of the University of Craiova, Proceedings of electrical engineering series, vol 35, pp 123–125. Annals of the University of Craiova, Electrical Engineering Series (2011)

Gong Y, Radachowsky SE, Wolf M, Nielsen ME, Girguis PR, Reimers CE (2011) Benthic microbial fuel cell as direct power source for an acoustic modem and seawater oxygen/temperature sensor system. Environ Sci Technol 45(11):5047–5053

Gutiérrez J, Villa-Medina JF, Nieto-Garibay A, Porta-Gándara MÁ (2014) Automated irrigation system using a wireless sensor network and GPRS module. IEEE Trans Instrum Meas 63(1):166–176

Jamain UM, Ibrahim NH, Ab Rahim R (2014) August. Performance analysis of zinc oxide piezoelectric MEMS energy harvester, In 2014 IEEE International Conference on Semiconductor Electronics (ICSE2014). IEEE, pp 263–266

Jha SK (1976) Characteristics and sources of noise and vibration and their control in motor cars. J Sound Vib 47(4):543–558

Karami MA, Bilgen O, Inman DJ, Friswell MI (2011) Experimental and analytical parametric study of single-crystal unimorph beams for vibrationenergy harvesting. IEEE transactions on ultrasonics, ferroelectrics, and frequency control 58(7):1508–1520

Knight C, Davidson J, Behrens S (2008) Energy options for wireless sensor nodes. Sensors 8(12):8037–8066

Koenig D, Chiaramont M, Balbinot A (2008) Wireless network for measurement of whole-body vibration. Sensors 8(5):3067–3081

Kulah H, Najafi K (2008) Energy scavenging from low-frequency vibrations by using frequency up-conversion for wireless sensor applications. IEEE Sens J 8(3):261–268

Li X, Hehn T, Thewes M, Kuehne I, Frey A, Scholl G, Manoli Y (2013) Non-resonant electromagnetic energy harvester for car-key applications. J Phys Conf Ser 476(1):012096

Liu Y (2012a) Wireless sensor network applications in smart grid: recent trends and challenges. Int J Distrib Sens Netw 8(9):492819

Liu C (2012b) Foundations of MEMS. Pearson Education India

Löhndorf, M., Kvisterøy, T., Westby, E. and Halvorsen, E., 2007. Evaluation of energy harvesting concepts for tire pressure monitoring systems. *Proceedings of Power MEMS*, pp. 331–334

Lossec M, Multon B, Ahmed HB (2013) Sizing optimization of a thermoelectric generator set with heatsink for harvesting human body heat. Energy Convers Manag 68:260–265

Martin P, Charbiwala Z, Srivastava M (2012, November) DoubleDip: leveraging thermoelectric harvesting for low power monitoring of sporadic water use. In: Proceedings of the 10th ACM conference on embedded network sensor systems, pp 225–238. ACM

Marzencki M, Defosseux M, Basrour S (2009) MEMS vibration energy harvesting devices with passive resonance frequency adaptation capability. J Microelectromech Syst 18(6):1444–1453

Morais R, Matos SG, Fernandes MA, Valente AL, Soares SF, Ferreira PJSG, Reis MJCS (2008) Sun, wind and water flow as energy supply for small stationary data acquisition platforms. Comput Electron Agric 64(2):120–132

Nicolson AM (1919) The piezo electric effect in the composite rochelle salt crystal. Trans Am Inst Electr Eng 38(2):1467–1493

O'Riordan E, Blokhina E, Galayko D (2016, December) Electromechanical coupling in electrostatic kinetic energy harvesters. In: 2016 IEEE international conference on electronics, circuits and systems (ICECS), pp 436–437. IEEE

Pietrelli A, Micangeli A, Ferrara V, Raffi A (2014) Wireless sensor network powered by a terrestrial microbial fuel cell as a sustainable land monitoring energy system. Sustainability 6(10):7263–7275

Reimers CE, Tender LM, Fertig S, Wang W (2001) Harvesting energy from the marine sediment—water interface. Environ Sci Technol 35(1):192–195

Rizzon L, Rossi M, Passerone R, Brunelli D (2013, November) Wireless sensor networks for environmental monitoring powered by microprocessors heat dissipation. In: Proceedings of the 1st international workshop on energy neutral sensing systems, p 8. ACM

Roundy S, Otis BP, Chee YH, Rabaey JM, Wright P (2003) A 1.9 GHz RF transmit beacon using environmentally scavenged energy. Optimization 4(2):4

Sample AP, Meyer DT, Smith JR (2011) Analysis, experimental results, and range adaptation of magnetically coupled resonators for wireless power transfer. IEEE Trans Industr Electron 58(2):544–554

Sardini E, Serpelloni M (2011) Self-powered wireless sensor for air temperature and velocity measurements with energy harvesting capability. IEEE Trans Instrum Meas 60(5):1838–1844

Sazonov E, Li H, Curry D, Pillay P (2009) Self-powered sensors for monitoring of highway bridges. IEEE Sens J 9(11):1422–1429

Scorcioni S, Bertacchini A, Dondi D, Larcher L, Pavan P, Mainardi G (2011, November) A vibration-powered wireless system to enhance safety in agricultural machinery. In: IECON 2011–37th annual conference of the IEEE industrial electronics society, pp 3510–3515. IEEE

Soliman MSM, El-Saadany EF, Mansour RR (2006, July) Electromagnetic MEMS based micro-power generator. In: 2006 IEEE international symposium on industrial electronics, vol 4, pp 2747–2753. IEEE

Sun H, Guo YX, He M, Zhong Z (2012) Design of a high-efficiency 2.45-GHz rectenna for low-input-power energy harvesting. IEEE Antennas Wirel Propag Lett 11:929–932

Szefi JT, Smith EC, Lesieutre GA (2004) Design and testing of a compact layered isolator for high-frequency helicopter gearbox isolation. In: Collection of technical papers-AIAA/ASME/ASCE/AHS/ASC structures, structural dynamics and materials conference, vol 6, pp 4596–4608. American Institute of Aeronautics and Astronautics Inc. (AIAA)

Tan YK, Panda SK (2011) Self-autonomous wireless sensor nodes with wind energy harvesting for remote sensing of wind-driven wildfire spread. IEEE Trans Instrum Meas 60(4):1367–1377

Tolentino IM, Talampas MR (2012, October) Design, development, and evaluation of a self-powered GPS tracking system for vehicle security. In: SENSORS, 2012 IEEE, pp 1–4. IEEE

Vijayaraghavan K, Rajamani R (2010) Novel batteryless wireless sensor for traffic-flow measurement. IEEE Trans Veh Technol 59(7):3249–3260

Vullers R, van Schaijk R, Gyselinckx B, Van Hoof C (2009, June) Is there a sweet spot for energy harvesting? In: 2009 device research conference, pp 7–8. IEEE

Williams CB, Yates RB (1996) Analysis of a micro-electric generator for microsystems. sensors and actuators A: Physical 52(1–3):8–11

Yerva L, Campbell B, Bansal A, Schmid T, Dutta P (2012, April) Grafting energy-harvesting leaves onto the sensornet tree. In: Proceedings of the 11th international conference on information processing in sensor networks, pp 197–208. ACM

Yuan Q, Sun X, Low CH, Fang DM, Zhang H (2010) Electromagnetic energy harveser fabricated with electrodeposition process. In: Proceedings of PowerMEMS 2010, pp 95–98

Zhang C, Syed A, Cho Y, Heidemann J (2011, November) Steam-powered sensing. In: Proceedings of the 9th ACM conference on embedded networked sensor systems, pp 204–217. ACM

Zhao C, Yisrael S, Smith JR, Patel SN (2014, September) Powering wireless sensor nodes with ambient temperature changes. In: Proceedings of the 2014 ACM international joint conference on pervasive and ubiquitous computing, pp 383–387. ACM

Zhu D, Tudor MJ, Beeby SP (2009) Strategies for increasing the operating frequency range of vibration energy harvesters: a review. Measurement Science and Technology 21(2):p 022001

Chapter 2
Piezoelectric Vibration Energy Harvesters: A Review

Abstract In this chapter, a state-of-the-art review focused on MEMS-based P-VEHs is presented. Different device designs based on single-beam structure fixed-free type, generator array type and fixed-fixed or guided beam type are presented. Key parameters like maximum power output, resonance frequency and fabrication technology used are discussed. Unimorph and bimorph piezoelectric layer-based designs operating in different modes are presented. Finally, the findings are summarized, and literature gaps based on low-frequency design and stable device operation are discussed.

Keywords Piezoelectric vibration energy harvester · Cantilever beam · Generator array · Guided beam

As discussed in the introduction, MEMS-based P-VEHs are suitable for WSNs.

A MEMS-based P-VEH can be realized by utilizing a cantilever structure free from one end and fixed at the other as shown in Fig. 2.1. Piezoelectric layer is sandwiched between the top and bottom electrodes. Displacement of the cantilever induces stress throughout the cantilever beam which is generally maximum in the proximity of the fixed beam end and generates charges in the piezoelectric layer. Charge generated gets collected through the electrodes. A seismic mass deployed at the free end effectively reduces the resonance frequency of the piezoelectric energy harvester.

William and Yates (1996), presented first vibration energy-based microelectric generator. Vibration energy harvester has been modeled as a spring–mass–damper system of second order and associated with a single degree of freedom (Fig. 1.2, Chap. 1). The relative displacement or mechanical strain in the system results from the transmission of vibrations by the inertial frame to m, the suspended inertial mass via k and b, the spring and damper, respectively. The device proposed in this paper generates power of 100 μW at 330 Hz and 1 μW at 70 Hz giving a displacement of 50 μm.

Glynne-Jones et al. (2001), developed a single-beam cantilever-type piezoelectric energy harvester. They used thick-film piezoelectric technology in the production of electrical power from environmental vibrations by developing an inertial generator. Steel substrate of 100 μm thickness was used on which piezoceramic thick film of

S. Saxena et al., *Design and Development of MEMS based Guided Beam Type Piezoelectric Energy Harvester*, Energy Systems in Electrical Engineering,
https://doi.org/10.1007/978-981-16-0606-9_2

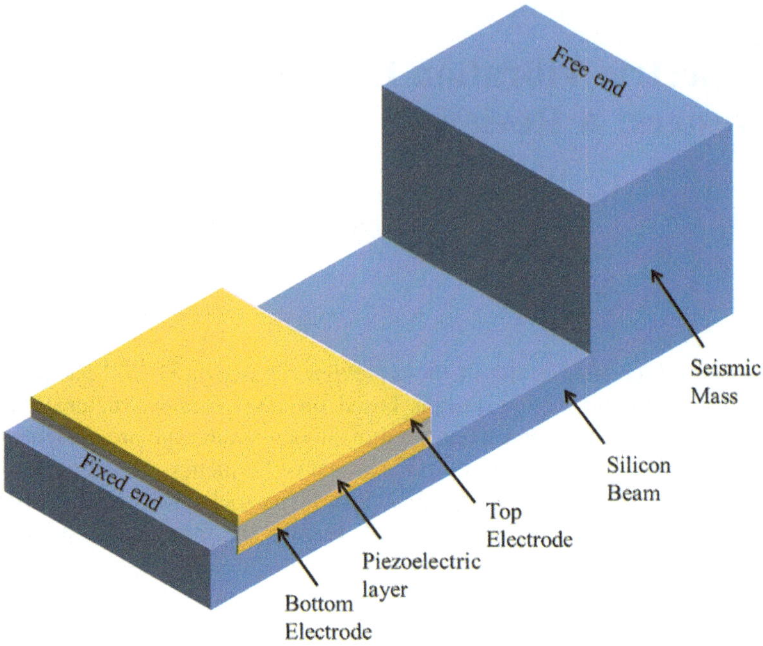

Fig. 2.1 Schematic of single-beam cantilever-type P-VEH

lead zirconate titanate (PZT) having a thickness of 70 μm has been deposited. The device has an experimental resonance frequency of 80.1 Hz and maximum output power delivering capability of 2.1 μW.

Shahruz (2006), designed an array of cantilevers having different length, width with different seismic mass at the end as shown in Fig. 2.2. As already discussed, a single-beam cantilever operates as a spring–mass–damper; therefore, the output at resonance is maximum and almost zero otherwise. Therefore, an array of cantilever that performs energy scavenging is deployed that acts as a band-pass filter. Individual cantilevers get excited for a particular frequency alone, so the array of cantilevers generates output for a range of frequencies thereby acting as a band-pass filter. The main advantage of this system is that peak output is maintained for a wider range of frequencies. The dimensions of the cantilevers have been selected in such a way that the resonance frequencies are close with each other giving maximum output power for wider frequency range. However, as the number of cantilevers is increased and only one cantilever operates at a time this system gives a low volume-to-output ratio.

Fang et al. (2006), developed a P-VEH using UV-LIGA technology. In this work, composite cantilever with nickel metal mass has been fabricated. A PZT layer of 1.64 μm is sandwiched between the bottom and top electrodes of titanium/platinum. The device resonates around 608 Hz and generates power output of 2.16 mW.

Marzencki et al. (2007), designed and fabricated P-VEH using AlN piezoelectric film in d31 mode for very low-amplitude vibrations. The device resonance frequency

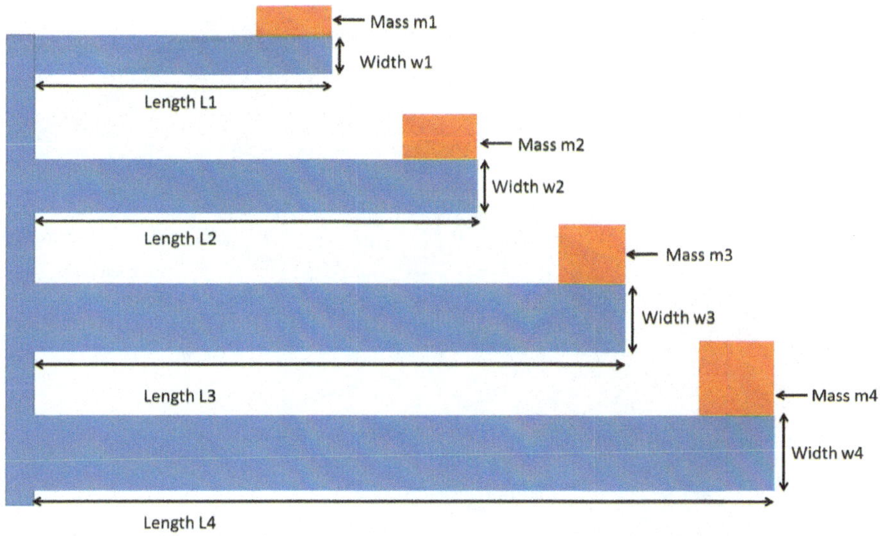

Fig. 2.2 An array of cantilever beams acting as band-pass filter: the beam dimensions and the proof masses at the tips are chosen in such a manner that the array acts as band-pass filter allowing a certain frequency range pass through

is found to be 214 Hz. When an input acceleration 0.126 g is introduced, the device generates a maximum power of 0.55 μW.

Xue et al. (2008), demonstrated an effective approach of integrating many piezoelectric biomorphs (PBs) in the design and implementation of a broadband piezoelectric energy harvester. Parallel and series connections of the bimorphs cantilevers were investigated. The series connection found to increase the frequency bandwidth of the system, whereas the parallel connection shifts the frequency bandwidth to the dominant frequency domain. Numerical analysis was performed on the two connections, and a mixed approach, i.e., combination of series and parallel bimorph cantilevers, was proposed for optimal output.

Muralt et al. (2009), fabricated and characterized a cantilever-based P-VEH having a rectangular-shaped seismic mass at its tip. A thick PZT (2 μm) layer was deposited on silicon substrate (5 μm) having interdigitated electrodes designed for higher voltage output. When excited with an input acceleration of 2 g, maximum output power of 1.4 μW along with maximum output voltage of 1.6 V has been achieved. The fabricated cantilever resonates at 870 Hz.

Lee et al. (2009), designed and fabricated P-VEH based on d33 and d31 modes. High-quality and efficient PZT thin films of thickness up to 28 μm have been deposited using aerosol deposition method on silicon cantilevers. At 2.5 g input acceleration, d31 mode gives an output power of 2.765 μW. The device resonance frequency is found to be 255.9 Hz. Similarly, the d33 mode gives a peak output power of 1.288 μW at 2 g input acceleration. Resonance frequency of the device reported here is 214 Hz.

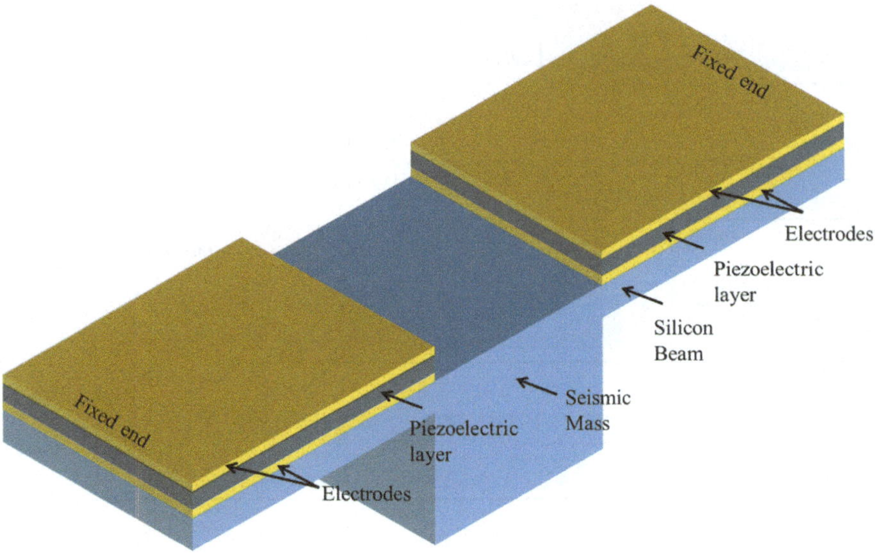

Fig. 2.3 Schematic of clamped–clamped or guided beam-type P-VEH

Marzencki et al. (2009), designed and developed clamped–clamped beam or guided two-beam structure where both the ends are fixed and have a seismic mass at the center. A schematic of a guided two-beam P-VEH is shown in Fig. 2.3. A clamped–clamped structure with two beams having width 2000 μm and length 1200 μm, a seismic mass having length width and thickness of 2200, 3000 and 450 μm, respectively, are connected at the center. The device has a resonance around 1800 Hz and shows a nonlinear behavior. The device nonlinearity is used for frequency adaptability from 1800 to 2450 Hz at 2 g. Around resonance, the device generates power of 8.269 nW at an input acceleration of 1 g. The clamped–clamped structure has a nonlinearity which results in frequency adaptability of 36%. Clamped–clamped beam structure is very stable and reliable in high-amplitude vibrations due to its ability to withstand higher stress; therefore, this structure is highly suitable for industrial applications where there is abundance of high-amplitude vibrations which the structure will have to withstand. Another, design of clamped-free beam-based device has seismic mass of length 800 μm and thickness of 450 μm. The beam length used is 400 μm. The resonance frequency of this structure is 1370 Hz. The fabricated device has 1-μm-thick AlN layer sputtered using dc-reactive magnetron sputtering. Aluminum top electrode is used for charge collection. At input acceleration of 1 g, the maximum voltage generated is 0.42 volts and maximum power generated is 0.1 μW.

Hirasawa et al. (2010), also developed P-VEH using AlN piezoelectric film. The device resonance frequency is 857 Hz, and when an input acceleration of 1 g is applied, it generates a maximum output of 0.18 μW.

Elfrink et al. (2010), designed and fabricated an AlN-based P-VEH and used vacuum packaging at the wafer level. Pt (Platinum) bottom electrode, sandwiched

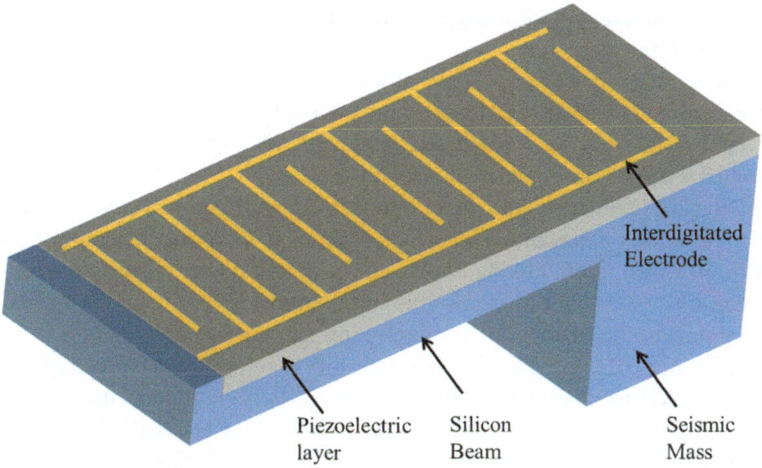

Fig. 2.4 Schematic of d33 mode cantilever-type P-VEH having interdigitated electrodes

AlN piezoelectric layer and Al (Aluminum) top electrodes were deposited and patterned. The device tested generates 17 μW of power at 0.64 g input acceleration and resonance frequency of 353 Hz. Packaged device generated 32 μW of maximum output power, whereas the same unpackaged device generated 10.2 μW of power. Due to vacuum packaging, the damping effects of air have been reduced which results in higher device output.

Blystad et al. (2010), designed a unimorph cantilever-type P-VEH for random vibrations. The base of the cantilever is of silicon, AlN is the piezoelectric layer and metal top electrode for collecting charge. Mechanical stoppers and power conditioning circuit were designed and simulated. It was observed that maximum power level varies when different power conditioning techniques were used. The mechanical stopper models used to affect the bandwidth of the system. The elastic stopper model used increases the bandwidth of the system. This paper designed and simulated cantilever structure with different power conditioning circuit with and without a mechanical stopper in a random vibration environment.

Park et al. (2010), presented fabrication as well as characterization of a P-VEH in d33 piezoelectric mode. A design schematic of d33 mode P-VEH having interdigitated (IDE) electrodes on the top of the piezoelectric layer is shown in Fig. 2.4. Here, thin films of PZT were deposited by sol-gel spin coating method. The interdigitated electrode structure was fabricated on the piezoelectric layer. The device generated a maximum output power of 1.1 μW when a load of 2.2 MΩ is considered for an input applied acceleration of 0.39 g. The device resonance frequency reported is 528 Hz. The device utilized interdigitated electrodes to harvest energy from d33 mode of the piezoelectric layer.

Morimoto et al. (2010), fabricated P-VEH by depositing PZT film on steel cantilevers ensuring higher energy conversion efficiency and robustness. Thin films of PZT were deposited on Pt/MgO substrates using RF-magnetron sputtering. The fabricated cantilever, without using a seismic mass at the cantilever tip, resonated at 126 Hz. At an input excitation acceleration of 5 m/s^2, the fabricated device generated an maximum output power of 5.3 μW at a load of 50 kΩ.

Durou et al. (2010), designed and developed a piezoelectric energy harvester with cantilever structure having a resonance frequency of 76 Hz. The device generated power of 13.9 μW at an input acceleration 0.2 g.

Hu et al. (2011), investigated the effect of strain gradient on a cantilever bimorph P-VEH. The strain on the cantilever structure tends to increase the natural frequency of the PB. However, it effectively improved the output power density of the P-VEH. It has been observed that the strain-gradient coupling coefficients tend to induce a frequency shift to the PB operation thereby influencing its performance.

Chang et al. (2011), designed a piezoelastic generator for harvesting vibration energy generated by computer hard disk drives. Polyvinylidene fluoride (PVDF) layer of 25 μm thickness was used as a piezoelectric layer. Harvester having different modes of deformation from few hertz to kilohertz was designed and fabricated. A maximum power of 0.55 W from the energy harvester has been achieved.

Karami et al. (2011), designed and developed unimorph cantilever-type P-VEH. A total of 24 unimorphs were fabricated using different PZT-5H, PZT-5A piezoceramics. The cantilevers have a resonance frequency from 500 to 600 Hz. The current output ranges from 0.1 to 1 A, and the maximum power extracted was 100 μW.

Liu et al. (2011), designed and fabricated a cantilever-based P-VEH for low-frequency operation. Parallel cantilevers are connected with a heavy silicon seismic mass at one end and fixed from the other. A thick 3 μm PZT layer is sandwiched between platinum/titanium electrodes. The device has an operation bandwidth of 17 Hz, i.e., from 30 to 47 Hz. Maximum power generation of 51.3 nW has been reported in this paper.

Hajati and Kim (2011), designed and fabricated doubly clamped beam structure that is fixed at both the ends and having a proof mass at the beam center. This design utilizes the stretching strain of the beams, therefore giving an improvement in the bandwidth and the power density. A PZT layer of 0.27 μm thickness is deposited over the beams. Thousand pairs of interdigitated electrodes are defined on top of the PZT layer, to harvest potential in d33 mode. At 1300 Hz, the fabricated device generates a maximum power of 22 μW across the optimized load resistance of 290 kΩ.

Lei et al. (2011), designed and developed a PZT-based P-VEH. The device resonates at 235 Hz. The maximum output was 14 μW when 1 g excitation is applied.

Defosseux et al. (2012), designed and fabricated a single-beam P-VEH for low-amplitude vibrations. The device resonates at 200 Hz when an acceleration lower than 0.25 g is applied. Initially, the 0.1 μm Mo bottom electrode was sputter deposited followed by which DC-pulsed reactive sputter deposition of 2-μm-thick AlN was done. A 0.1-μm-thick sputter deposited Pt forms the top electrode. At input excitation of 0.25 g, the developed device harvests a peak power of 0.62 μW and resonates at a frequency of 214 Hz.

Xu et al. (2012), presented a PZT thick-film PB (with two piezoelectric layers) P-VEH having a proof mass made of silicon. A schematic of a bimorph cantilever-type P-VEH is shown in Fig. 2.5. The coupling coefficient of the system was seen

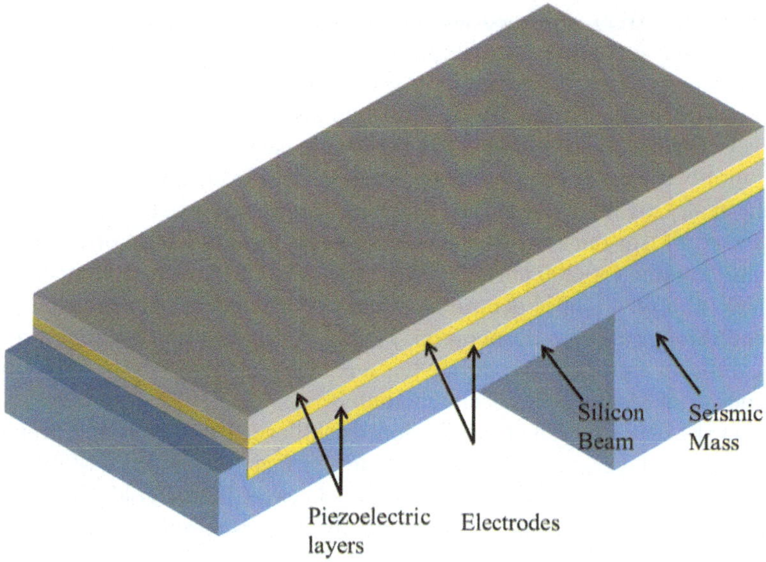

Fig. 2.5 Schematic of a bimorph cantilever-type P-VEH

to increase due to presence of two piezoelectric layers, and thus, the output power density increases significantly. At an input acceleration of 1 g, the harvester has maximum output power of 37.1 μW.

Kanno et al. (2012), designed and fabricated cantilever-based P-VEH using different piezoelectric materials. Thin films of Pb-free sodium potassium niobate $(K,Na)NbO_3$ (KNN) and $Pb(Zr,Ti)O_3$ (PZT) were deposited on platinum/titanium/silicon substrates via RF-magnetron sputtering. The maximum output power of the device with KNN/Si is 1.1 μW and that of PZT/Si is 1.0 μW. This paper compared the two piezoelectric materials KNN and PZT and found that both devices have around equal power output. Therefore, a lead-free material KNN can be used which generated equal output without having lead contamination.

Al Ahmad and Jabbour (2012), designed and developed a cantilever-based P-VEH for free-falling droplets. The harvester is a fixed-free end cantilever having five layers of PZT stacked on top of one another. The cantilever has a resonance frequency of 315.15 Hz. For a moderate rain intensity of 230 drops per second, an energy generation of 400 μJoule has been reported.

Wang et al. (2012a), designed and fabricated an optimized clamped–clamped beam P-VEH (Fig. 2.3) for lower frequency vibrations. The mass beam length is 3780 μm, and the truss beam length is also 3780 μm. Pt, AlN and Al layers are sputter deposited sequentially and are patterned in the form of capacitors. DRIE is used to define the two beams and for releasing the structure. The fabricated device resonates at 694 Hz. When an acceleration of 1.2 g is used for excitation, the maximum output power generated by the structure is 10 μW from one set of capacitor. The device exhibits a tuning range of 3 Hz.

Sang et al. (2012), designed and developed cantilever-based hybrid vibration energy harvester for WSN. A cantilever beam having two piezoelectric layers with a magnet place at the tip of the cantilever is designed. The magnets on the tip and the bottom electrodes accelerate the vibration and also adjust the cantilever resonance frequency. The resonance frequency of the energy harvester is 50 Hz. At 0.4 g input acceleration, 10.7 mW power has been generated at a load end of 50 Ω.

Wang et al. (2012b), designed and developed cantilever-type P-VEH having metal substrate. A 50-μm-thick layer of stainless steel is used as a base of the cantilever. Titanium/Platinum (30/200 nm) films were sputtered on steel for bottom electrodes. PZT thick film of 4 μm is deposited using sol-gel process. Al top electrode was deposited through e-beam evaporation. Finally, a copper proof mass gets attached at the cantilever free end. The device has a resonating frequency of 89 Hz, and an output power of 15.4 μW has been reported.

Tang et al. (2012), designed and developed cantilever-based P-VEH. Chrome/gold (20 nm/150 nm) layers were deposited to form primary electrodes. PMN-PT piezoelectric layer was deposited having 10 mm thickness. Chrome/gold was again sputtered to form the top electrode. A nickel-seismic mass is introduced at the cantilever beam tip. Resonance frequency of 237.4 Hz and a maximum power generation of 2.704 μW have been reported.

Berdy et al. (2012), designed and developed P-VEH. PZT layer is sandwiched between nickel electrodes. The fabricated device resonates at 50 Hz. Maximum power density of 105 μW at load resistance (380 KΩ) end has been reported in this paper.

Elvin and Elvin (2013), investigated vibration energy harvesting due to human motion suitable for wireless implants. Human foot motion was deeply investigated using multi-axis accelerometer giving the displacement and the frequency. For a device that can operate for a high acceleration of 30 g would generate a power of 28 mW. Effects of electromechanical coupling coefficient and mechanical tuning have been investigated in this paper. A higher coupling coefficient and tuning result in increased output power.

Yun and Yun (2013), designed and developed a woven piezoelectric energy harvester for wireless portable devices. Polymer threads were used in the column, and the piezoelectric film weaving them in the row was designed. Due to stretching and contracting, motion charges are generated and collected through polymer patches. At an operation frequency of 8 Hz due to continuous contraction and stretching, a peak power generation of 0.63 mW/cm^2 has been reported.

Gao et al. (2013), designed and developed cantilever-based flow energy harvester. PZT layer 127 μm thick is used in designing the energy harvester. The lowest resonance frequency achieved was 20.7 Hz. A change in flow generates a voltage fluctuation of 1 V which has been reported in the paper.

Usman et al. (2014), designed and developed energy harvester for acquiring, processing and transmitting the electrocardiogram (ECG) data in a WSN. A cantilever structure having seismic mass at the end is deployed to harvest energy. PZT layer is sandwiched between platinum electrodes. The proposed design when integrated into a system can perform consuming only few μWs of power from heartbeats having a resonance frequency 39 Hz and acceleration 0.3 m/s^2.

Alsuwaiyan et al. (2014), presented an approach to harvest energy from low-frequency vibrations. Three cantilever-type P-VEHs having different resonance frequency of 40, 45, 53.4 Hz have been used. The maximum power generation of 440 μW has been reported in this paper.

Xu and Kim (2015), designed and fabricated clamped–clamped beam device for harvesting energy from low-frequency and low-amplitude vibrations. The device testing done shows operating frequency in the range 50–150 Hz at an input acceleration of 0.2 g. Both macro-prototype and MEMS device have been presented in this paper.

Saadon and Wahab (2015), designed a cantilever array for low-frequency operation range of 67–70 Hz. A PZT layer of 1.2 μm thickness is sandwiched between 0.1-μm-thick platinum electrodes. Input acceleration range of 0.2–1.3 g maximum power generation of 6.8 μW has been reported in this paper.

Emamian et al. (2016), designed and developed a piezoelectric energy harvester using a screen printing technique. The structure consists of a thick PVDF layer of 12.5 μm sandwiched between METglas electrodes. The device has a first resonance frequency of 54 Hz. Across a load of 1 MΩ maximum power generation of 0.28 μW has been reported in the paper.

Nguyen et al. (2017),designed and developed AlN-based energy harvester on titanium foils. A 4 μm AlN thick film is deposited over titanium foil. Chrome/gold was deposited to form top electrodes. The fabricated device resonates at a frequency of 645.3 Hz. Maximum harvested power of 13.1 μW has been reported in this paper.

Wang et al. (2017), designed and characterized a double-fork P-VEH. The designed harvester has first three resonance frequencies of 21, 47 and 91 Hz, respectively. The bandwidth of the designed energy harvester is 8 Hz at an input acceleration of 1 g.

Jackson et al. (2018), designed and developed a piezoelectric energy harvester for powering a pacemaker. AlN of thickness 0.45 μm is sandwiched between titanium and aluminum to form bottom and top electrodes. The device has a resonance at 28.5 and 678 Hz. Maximum power of 17 μW at 678 Hz has been reported in this paper.

Wen et al. (2018), designed and developed P-VEH for wind energy. Cantilever-based structure is designed which is actuated using magnets. The dimensions of the cantilever are length 100 mm and a thickness of 0.4 mm. The total power generated from the energy harvester at 360 rpm is 1.38 μW.

Nabavi and Zhang (2019), designed and developed a nonlinear wideband P-VEH. The designed structure has a doubly clamped beam with three seismic mass. The device has a wideband operation of 59 Hz from 227 to 286 Hz. The power density of the device reported is 595.12 $\mu W\ cm^3\ m^{-2}\ s^4$.

Wen and Xu (2019), designed and developed a novel integrated multistage energy harvester. Multi-objective genetic algorithm has been used to optimize the device dimensions. Across a matched load, the device exhibits peak power of 50.8 mW.

2.1 Discussion and Planning

The literature survey done states the suitability of cantilever-based P-VEH for WSNs. From the survey, the P-VEH can be broadly classified into three types on the basis of their structures: (i) single-beam cantilever unimorph or bimorph structures free form one end and fixed at the other, (ii) cantilever arrays with multiple cantilever elements connected in series or parallel and (iii) clamped–clamped/guided beam structures fixed at both the ends and having seismic mass attached at the center. The single-beam cantilever structure has low resonance frequency and gives high output because of large displacement of the beam; however, large displacement makes device unstable and results in beam fractures. Cantilever array-type structure gives peak output for wider frequency range due to large number of cantilevers; however, only one cantilever operates at a time results in low volume-to-output ratio. Guided beam structure gives reliable and stable device operation due to the fixed nature of beams; however, fixed beams increase the resonance frequency of the structure. Considering high-amplitude vibrations, guided beam structure-type P-VEH is a preferred choice due to stable operation and suitability of fabrication in MEMS technology. Therefore, we have focused our work on guided beam-type P-VEH. While doing literature survey, the following literature gaps were observed on guided beam-type P-VEH.

(a) Resonance frequency reported by Marzencki et al. (2009) for guided two-beam structure is 1800 and 694 Hz reported by Wang et al. (2012a). Considering low-frequency vibrations below 500 Hz, the resonance frequency of the reported devices is relatively higher. Therefore, there is a need to optimize the fabrication technology in order to further lower the device resonance frequency.
(b) Guided two-beam structure utilizes a pair of split electrodes which harvests the tensile and compressive stress (Marzencki et al. 2009). However, in previous designs, the split electrode is deployed on complete or partial beams without optimizing the dimensions of the electrodes. Therefore, optimization of dimensions of split electrodes is necessary for maximum output with the lowest resonance frequency.
(c) Shape of the seismic mass is generally square on shape because it can be easily realized in MEMS technology using DRIE. Investigation on different shapes of seismic mass such as triangular shape and pyramidal shape has not been investigated. Therefore, effect of different shapes of seismic mass on potential generated by P-VEH needs to be investigated.
(d) Design and development of guided two-beam P-VEH have been reported (Marzencki et al. 2009; Wang et al. 2012a); however, design and fabrication of a guided four-beam P-VEH have not been reported so far. Therefore, design, fabrication of four-beam P-VEH need to be investigated and have to be compared with its guided two-beam counterpart.
(e) Testing of the two-beam and four-beam P-VEH in very low frequency up to 10 Hz has not been reported as devices have higher resonance frequency. Therefore, low-frequency testing of the devices has to be investigated experimentally to test the behavior of the devices in low-frequency domain.

As it is clear from the theory of a piezoelectric vibration energy harvester, when a cantilever beam undergoes a vibration in tune with the ambient vibrational energy, it develops a stress in its both the surfaces, and if both or one of its surfaces are coated with a piezoelectric (PE) layer, this PE layer will also be stressed along with the surface and acquires an electric charge. In this way, electric charge is generated as a result of ambient vibration energy. It will generate a voltage if PE layer is in a closed electrical loop with a load. This activity of generating charge or voltage in a PE layer on top surface of a cantilever depends on the stiffness of the cantilever. Lower is the stiffness; higher will be the charge or voltage generation, i.e., higher sensitivity. It can be further improved by putting a proof mass at the free tip of the cantilever, but it makes cantilever fragile (Marzencki et al. 2009). In order to have a robust structure, we have chosen a beam with the hanging mass at its center. This is called a two-beam guided structure. The structure will be even more stable if we take four beams at right angles to each other with a common suspended mass at their tip. This is called a four-beam guided structure. In this work, we have taken these two structures for the design of our vibration energy harvester. Therefore, taking the above research gaps into consideration, we aim at design and development of guided two-beam and guided four-beam P-VEH for low-frequency operation.

References

Al Ahmad M, Jabbour GE (2012) Electronically droplet energy harvesting using piezoelectric cantilevers. Electron Lett 48(11):647–649

Alsuwaiyan AS, Eltaib MEH, Sherif HA (2014, August) Ambient vibrations piezoelectric harvester array with discrete multiple low frequencies. In: 2014 2nd international conference on technology, informatics, management, engineering & environment, pp 174–178. IEEE

Berdy DF, Srisungsitthisunti P, Jung B, Xu X, Rhoads JF, Peroulis D (2012) Low-frequency meandering piezoelectric vibration energy harvester. IEEE Trans Ultrason Ferroelectr Freq Control 59(5):846–858

Blystad LCJ, Halvorsen E, Husa S (2010) Piezoelectric MEMS energy harvesting systems driven by harmonic and random vibrations. IEEE Trans Ultrason Ferroelectr Freq Control 57(4):908–919

Chang JY (2011) Modeling and analysis of piezo-elastica energy harvester in computer hard disk drives. IEEE Trans Magn 47(7):1862–1867

Defosseux M, Allain M, Defay E, Basrour S (2012) Highly efficient piezoelectric micro harvester for low level of acceleration fabricated with a CMOS compatible process. Sens Actuators, A 188:489–494

Durou H, Ardila-Rodriguez GA, Ramond A, Dollat X, Rossi C, Esteve D (2010) Micromachined bulk PZT piezoelectric vibration harvester to improve effectiveness over low amplitude and low frequency vibrations. In: Proceedings of Power MEMS'10, pp 27–30

Elfrink R, Renaud M, Kamel TM, De Nooijer C, Jambunathan M, Goedbloed M, Hohlfeld D, Matova S, Pop V, Caballero L, Van Schaijk R (2010) Vacuum-packaged piezoelectric vibration energy harvesters: damping contributions and autonomy for a wireless sensor system. J Micromech Microeng 20(10):104001

Elvin NG, Elvin AA (2013) Vibrational energy harvesting from human gait. IEEE/ASME Trans Mechatron 18(2):637–644

Emamian S, Chlaihawi AA, Narakathu BB, Bazuin BJ, Atashbar MZ (2016, October) A piezoelectric based vibration energy harvester fabricated using screen printing technique. In: 2016 IEEE Sensors, pp 1–3. IEEE

Fang HB, Liu JQ, Xu ZY, Dong L, Wang L, Chen D, Cai BC, Liu Y (2006) Fabrication and performance of MEMS-based piezoelectric power generator for vibration energy harvesting. Microelectron J 37(11):1280–1284

Gao X, Shih WH, Shih WY (2013) Flow energy harvesting using piezoelectric cantilevers with cylindrical extension. IEEE Trans Industr Electron 60(3):1116–1118

Glynne-Jones P, Beeby SP, White NM (2001) Towards a piezoelectric vibration-powered microgenerator. IEE Proc Sci Measur Technol 148(2):68–72

Hajati A, Kim SG (2011) Ultra-wide bandwidth piezoelectric energy harvesting. Appl Phys Lett 99(8):083105

Hirasawa TH, Yen YTT, Wright PK, Pisano AP, Lin L (2010) Design and fabrication of piezoelectric aluminum nitride corrugated beam energy harvester. In: Proceedings of Power MEMS, pp 211–214

Hu Y, Wang J, Yang F, Xue H, Hu H, Wang J (2011) The effects of first-order strain gradient in micro piezoelectric-bimorph power harvesters. IEEE Trans Ultrason Ferroelectr Freq Control 58(4):849–852

Jackson N, Olszewski O, Mathewson A, O'Murchu C (2018, January) Ultra-low frequency Piezo-MEMS energy harvester for a leadless pacemaker. In: 2018 IEEE micro electro mechanical systems (MEMS), pp 642–645. IEEE

Kanno I, Ichida T, Adachi K, Kotera H, Shibata K, Mishima T (2012) Power-generation performance of lead-free (K, Na) NbO3 piezoelectric thin-film energy harvesters. Sens Actuat, A 179:132–136

Karami MA, Bilgen O, Inman DJ, Friswell MI (2011) Experimental and analytical parametric study of single-crystal unimorph beams for vibration energy harvesting. IEEE Trans Ultrason Ferroelectr Freq Control 58(7):1508–1520

Lee BS, Lin SC, Wu WJ, Wang XY, Chang PZ, Lee CK (2009) Piezoelectric MEMS generators fabricated with an aerosol deposition PZT thin film. J Micromech Microeng 19(6):065014

Lei A, Xu R, Thyssen A, Stoot AC, Christiansen TL, Hansen K, Lou-Moeller R, Thomsen EV, Birkelund K (2011, January) MEMS-based thick film PZT vibrational energy harvester. In: 2011 IEEE 24th international conference on micro electro mechanical systems, pp 125–128. IEEE

Liu H, Tay CJ, Quan C, Kobayashi T, Lee C (2011) Piezoelectric MEMS energy harvester for low-frequency vibrations with wideband operation range and steadily increased output power. J Microelectromech Syst 20(5):1131–1142

Marzencki M, Basrour S, Belgacem B, Muralt P, Colin M (2007, May) Comparison of piezoelectric MEMS mechanical vibration energy scavengers. In: Nanotech 2007, pp 21–24. CRC Press

Marzencki M, Defosseux M, Basrour S (2009) MEMS vibration energy harvesting devices with passive resonance frequency adaptation capability. J Microelectromech Syst 18(6):1444–1453

Morimoto K, Kanno I, Wasa K, Kotera H (2010) High-efficiency piezoelectric energy harvesters of c-axis-oriented epitaxial PZT films transferred onto stainless steel cantilevers. Sens Actuators, A 163(1):428–432

Muralt P, Marzencki M, Belgacem B, Calame F, Basrour S (2009) Vibration energy harvesting with PZT micro device. Procedia Chem 1(1):1191–1194

Nabavi S, Zhang L (2019) Nonlinear multi-mode wideband piezoelectric MEMS vibration energy harvester. In: IEEE Sensors Journal

Nguyen MS, Yoon YJ, Kwon O, Kim P (2017) Lowering the potential barrier of a bistable energy harvester with mechanically rectified motion of an auxiliary magnet oscillator. Appl Phys Lett 111(25):253905

Park JC, Park JY, Lee YP (2010) Modeling and characterization of piezoelectric d_33-mode MEMS energy harvester. J Microelectromech Syst 19(5):1215–1222

Saadon S, Wahab Y (2015, November) From ambient vibrations to green energy source: MEMS piezoelectric energy harvester for low frequency application. In: 2015 IEEE student symposium in biomedical engineering & sciences (ISSBES), pp 59–63. IEEE

Sang Y, Huang X, Liu H, Jin P (2012) A vibration-based hybrid energy harvester for wireless sensor systems. IEEE Trans Magn 48(11):4495–4498

Shahruz SM (2006) Design of mechanical band-pass filters for energy scavenging. J Sound Vib 292(3–5):987–998

Tang G, Liu J-Q, Yang B, Luo J-B, Liu H-S, Li Y-G, Yang C-S, Dao V-D, Tanaka K, Sugiyama S (2012) Piezoelectric MEMS low-level vibration energy harvester with PMN-PT single crystal cantilever. Electron Lett 48(13): 784–786. https://doi.org/10.1049/el.2012.1419

Usman A, Mukhtar M, Umer RM (2014, December) MEMS based wireless energy harvesting mechanism for ECG and BCG application. In: 2014 International Conference on Emerging Technologies (ICET), pp 142–146. IEEE

Wang Z, Matova S, Elfrink R, Jambunathan M, De Nooijer C, van Schaijk R, Vullers RJM (2012a, January) A piezoelectric vibration harvester based on clamped-guided beams. In: 2012 IEEE 25th International Conference on Micro Electro Mechanical Systems (MEMS), pp 1201–1204. IEEE

Wang Q, Cao ZP, Kuwano H (2012b) Metal-based piezoelectric energy harvesters by direct deposition of PZT thick films on stainless steel. Micro Nano Lett 7(12):1158–1161

Wang X, Chen C, Wang N, San H, Yu Y, Halvorsen E, Chen X (2017) A frequency and bandwidth tunable piezoelectric vibration energy harvester using multiple nonlinear techniques. Appl Energy 190:368–375

Wen S, Xu Q (2019) Design of a novel piezoelectric energy harvester based on integrated multi-stage force amplification frame. IEEE/ASME Transactions on Mechatronics

Wen S, Xu Q, Zi B (2018) Design of a new piezoelectric energy harvester based on compound two-stage force amplification frame. IEEE Sens J 18(10):3989–4000

Williams CB, Yates RB (1996) Analysis of a micro-electric generator for microsystems. Sens Actuat A: Phys 52(1–3):8–11

Xu R, Kim SG (2015) Low-frequency, low-G MEMS piezoelectric energy harvester. In: Journal of Physics: Conference Series, vol 660, no 1, p 012013. IOP Publishing

Xu R, Lei A, Dahl-Petersen C, Hansen K, Guizzetti M, Birkelund K, Thomsen EV, Hansen O (2012) Fabrication and characterization of MEMS-based PZT/PZT bimorph thick film vibration energy harvesters. J Micromech Microeng 22(9):094007

Xue H, Hu Y, Wang QM (2008) Broadband piezoelectric energy harvesting devices using multiple bimorphs with different operating frequencies. IEEE Trans Ultrason Ferroelectr Freq Control 55(9):2104–2108

Yun D, Yun KS (2013) Woven piezoelectric structure for stretchable energy harvester. Electron Lett 49(1):65–66

Chapter 3
Design, Modeling and Comparison of Piezoelectric Vibration Energy Harvesters

Abstract In this chapter, analytical design and FEM analysis for cantilever-based P-VEH have been presented. Analytical equations giving device displacement, spring constant, resonance frequency for different structures have been reported. Device performance parameters such as resonance frequency, displacement, electric potential and von Mises stress are obtained using FEM for single-beam, cantilever array and guided-beam-type P-VEH. Effect of shape of seismic mass on the potential generated by the P-VEH has been investigated. Square-shape, pyramidal-shape and triangular-shape seismic mass-based cantilever structures have been designed, and displacement, stress and electric potential are obtained using FEM. Design giving better electric potential and stable response is selected for further investigation and device fabrication.

Keywords FEM · Cantilever beam · Piezoelectric · Seismic mass · Pyramidal Shape

Piezoelectric vibration energy harvesters give a promising solution to power WSNs by harvesting adequate power from ambient vibrations over a wide frequency range. P-VEHs are realized using a cantilever beam structure fixed from one end and free from the other. Majorly three types of structures are as follows: single-beam cantilever structure, cantilever array structure and guided beam structure are used in the development of MEMS-based P-VEHs. In this chapter, analytical design and FEM simulation of these three types of structures have been reported. Device parameters such as resonance frequency, cantilever displacement, output electric potential, maximum stress, stable operation and device reliability have been obtained and compared. Based on the comparison and ease of fabrication, a stable and reliable structure is selected and further investigated for device fabrication.

3.1 Single-Beam Cantilever-Type P-VEH

A single-beam cantilever structure fixed from one end and free from the other is used as a piezoelectric vibration energy harvester. A piezoelectric layer is sandwiched

S. Saxena et al., *Design and Development of MEMS based Guided Beam Type Piezoelectric Energy Harvester*, Energy Systems in Electrical Engineering,
https://doi.org/10.1007/978-981-16-0606-9_3

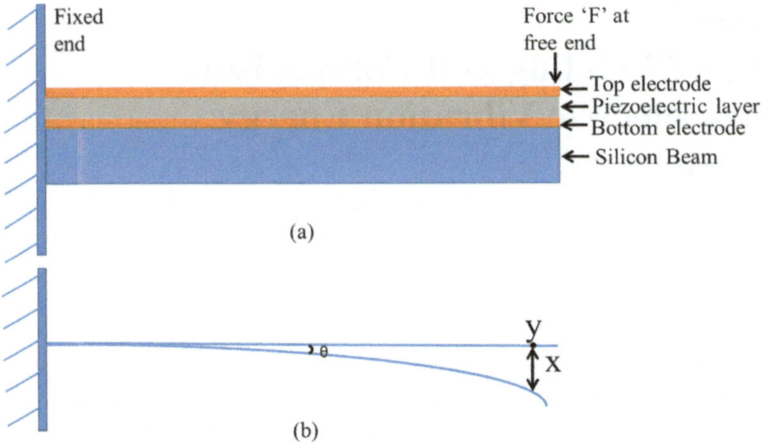

Fig. 3.1 **a** Single-beam cantilever and **b** beam deflection

between bottom and top electrodes as shown in Fig. 3.1a. Due to external vibrations, a force 'F' is exerted at the free end of the cantilever which results in the deflection of the beam from the free end making a curvature angle 'θ' as shown in Fig. 3.1b. The curvature of the beam due to small displacement can be obtained by solving a second-order differential equation given in Eq. 3.1 (Bao 2005).

$$\mathrm{EI}\frac{\mathrm{d}^2 y}{\mathrm{d}x^2} = M(x) \tag{3.1}$$

$M(x)$ is the bending moment of the beam cross section at position y, whereas x is the displacement at location y (Fig. 3.1b). 'E' is the Young's modulus of elasticity of the cantilever beam (measured in Pascal). 'I' is the moment of inertia.

At applied force 'F' (in Newton), the curvature angle 'θ' of the beam is given by Eq. 3.2 (Bao 2005), where 'l' is the length of the cantilever beam in μm.

$$\theta = \frac{\mathrm{Fl}^2}{2\mathrm{EI}} \tag{3.2}$$

The beam deflection of the beam or beam displacement 'x' is given by Eq. 3.3 (Bao 2005).

$$x = \frac{\mathrm{Fl}^3}{3\mathrm{EI}} \tag{3.3}$$

Rectangular-shaped cantilevers having length 'L' width 'w' and thickness 't' and the moment of inertia (I) with respect to neutral axis given by Eq. 3.4 are the most common type of structures used as P-VEH.

$$I = \frac{wt^3}{12} \tag{3.4}$$

MEMS-based cantilever structures are modeled as spring–mass systems. Considering a cantilever as a spring–mass system, the general expression of a force or spring constant is the force divided by the displacement as shown in Fig. 3.2.

The spring constant of the cantilever beam can be expressed in Eq. 3.5 (Bao 2005) using 3.3 and 3.4.

$$k = \frac{F}{x} = \frac{3EI}{l^3} = \frac{Ewt^3}{4l^3} \tag{3.5}$$

A cantilever beam is modeled as a spring–mass–damper system; therefore, the structure will have a fixed resonance frequency. Around this resonance frequency, the

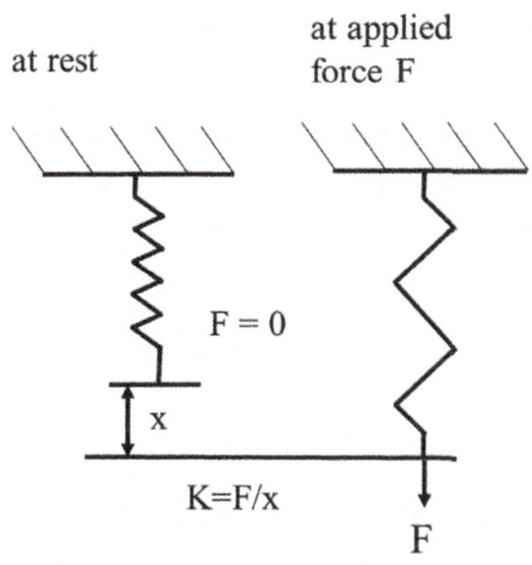

Fig. 3.2 Spring deformation at point load force 'F' (Williams and Yates 1996)

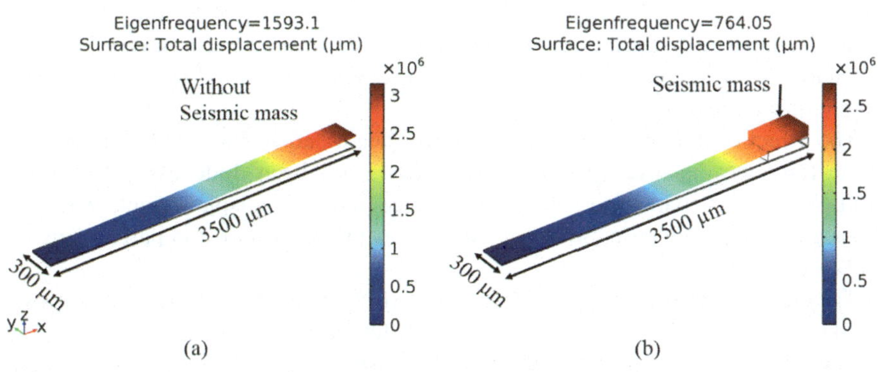

Fig. 3.3 **a** Resonance frequency of the free end cantilever is 1593.1 Hz. **b** Resonance frequency of the cantilever with proof mass is 764.05 Hz

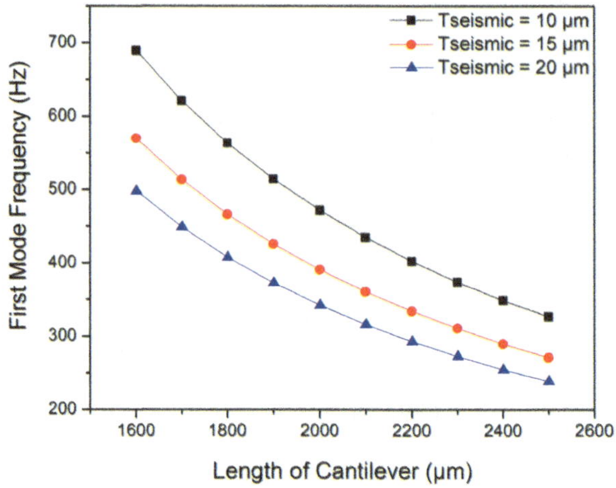

Fig. 3.4 Graph of resonance frequency versus cantilever length

cantilever beam will have a maximum displacement which is desired for maximum output. The resonance frequency 'f' of the cantilever depends on spring constant 'k' of the beam and the effective mass of the cantilever 'm' as given by Eq. 3.6 (Zhu et al. 2009).

$$f = \frac{1}{2\pi}\sqrt{\frac{k}{m}} \tag{3.6}$$

Ambient vibrations freely available in the environment are of low frequency range up to 2 kHz. Cantilever-type P-VEH must operate in resonance with low-frequency ambient vibrations for optimal output. Therefore, design parameters' length, width and thickness of the cantilever should be chosen such that to have the least possible resonance frequency. A seismic mass is placed at the tip of the cantilever to further reduce the resonance frequency. The effective mass 'm' will be the addition of cantilever mass and seismic mass. However, as the seismic mass is very heavy so the cantilever mass can be neglected. Here, the effective mass 'm' would be only the weight of the seismic mass.

The effect of dimensions of the cantilever beam and seismic mass on the resonance frequency is investigated. FEM analysis has been done using COMSOL Multiphysics® and parameters such as resonance frequency, displacement, von Mises stress and output electric potential have been obtained and compared with the analytical results.

The effect of seismic mass on the resonance frequency of the cantilever is investigated using COMSOL Multiphysics®. Two different cantilevers were designed one without seismic mass and the other with a seismic mass of dimensions 300 µm × 500 µm × 100 µm. The length, width and thickness of both the cantilever are 3500 µm, 300 µm and 10 µm, respectively. The eigenfrequency analysis is performed to obtain the resonance frequency of the cantilevers, and the results are shown in Fig. 3.3.

It can be seen from Fig. 3.3 that the free end cantilever without proof mass resonates at 1593.1 Hz, whereas when a seismic mass is placed at the free end of the cantilever, then the resonance frequency reduces to 764.05 Hz. Due to seismic mass, the resonance frequency reduces by 829 Hz or by 52% which is required to tune with low-frequency ambient vibrations. This is because the effective mass of the cantilever increases due to seismic mass. This statement also justifies Eq. 3.6. Another method to reduce the resonance frequency is to increase the length of the cantilever. As length increases, spring constant k decreases which reduces the resonance frequency of the cantilever as seen from Eq. 3.4.

Therefore, the effect of variation of length of the cantilever on resonance frequency is also investigated. After initial iterations, length variation from 1600 to 2500 μm is selected as it gives the desired frequency range from 200 to 700 Hz. Three different thicknesses of seismic mass 10, 15 and 20 μm are selected to reduce the resonance frequency.

The height of the seismic mass is selected from 10 to 20 μm because it can be easily electroplated during device fabrication. Figure 3.4 shows the effect of increase in length and seismic mass on the resonance frequency of the cantilever structure. As length is increased from 1600 to 2500 μm for the seismic mass of 10 μm thickness, the resonance frequency reduces from 689 to 326 Hz, and when seismic mass thickness is 20 μm, the resonance frequency reduces from 498 to 239 Hz. This can also be seen from the graph shown in Fig. 3.4 which also justifies Eqs. (3.4) and (3.5). Therefore, the two main parameters that decide resonance frequency of the cantilever-type energy harvester are the seismic mass '*m*' and the length of the cantilever 'l'. These two parameters can be varied in order to tune with the ambient vibrations.

To further investigate the nature of displacement, stress and magnitude of electric potential generated, a cantilever structure having dimensions as length 2000 μm, width 400 μm thickness 10 μm with a proof mass at the end of 400 μm × 400 μm × 20 μm is attached, and the structure is simulated and result is shown in Fig. 3.5.

Total displacement graph of the single cantilever structure is shown in Fig. 3.6. It can be seen that at the fixed end, i.e., 0 μm the displacement is minimum because the cantilever is fixed from that end and as we move toward the free end the displacement increases, maximum displacement is 8.5 μm at the free end.

Stress along the length of the cantilever is generated from fixed end toward free end along the *x*-axis. The region of maximum displacement has minimum stress due to free movement of cantilever, whereas at the fixed end the stress is maximum due to minimum displacement. Figure 3.7 gives the distribution of von Mises stress along the cantilever length. At the fixed end, the stress equals 6.5×10^6 N/m^2 and is maximum which reduces to minimum value as we move toward free end.

Electric potential is generated along the piezoelectric layer due to stress along the length of the cantilever length. Maximum stress at the fixed end generates maximum electric potential whereas minimum stress at free end results in minimum electric potential. Figure 3.8 gives the distribution of the electric potential along the length of the piezoelectric layer. At the fixed end, the electric potential is equal to 0.6 V

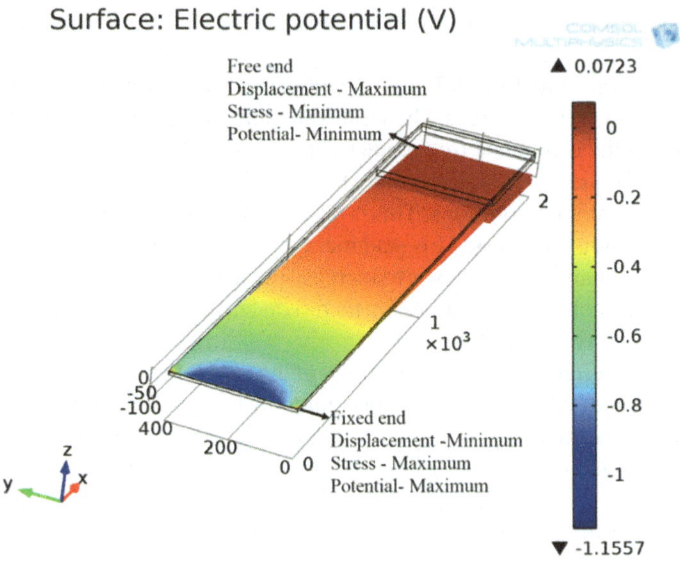

Fig. 3.5 FEM simulation of single-beam cantilever

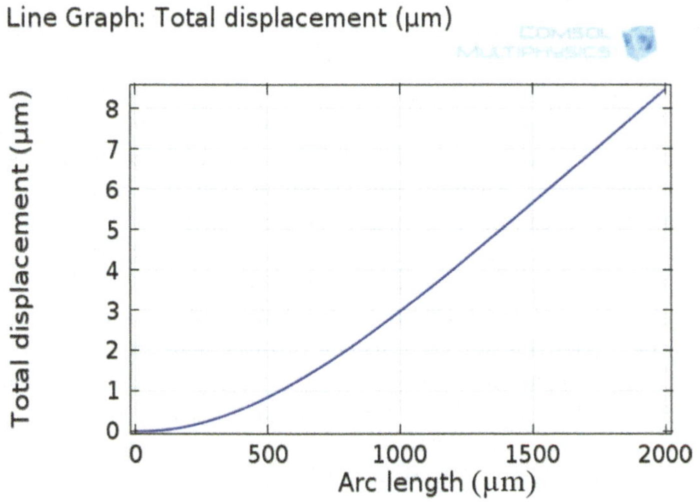

Fig. 3.6 Displacement along the length of the cantilever

and is maximum and (negative sign for polarity) which reduces as we move toward the free end of the cantilever. The electric potential will be maximum where there is maximum stress.

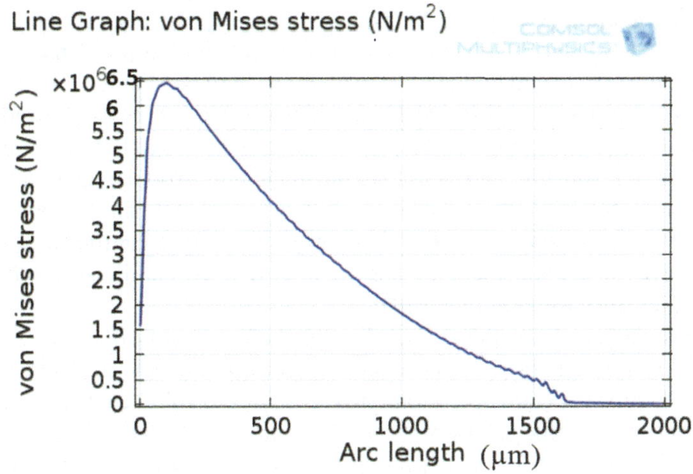

Fig. 3.7 von Mises stress along the length of the cantilever

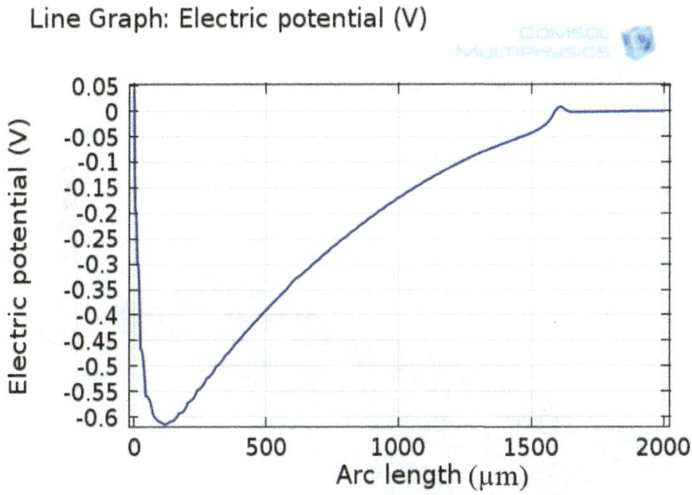

Fig. 3.8 Line graph of electric potential generated from fixed end toward free end

3.2 Cantilever Generator Array

A cantilever-type P-VEH behaves as a spring–mass–damper system, i.e., resonating only at a single frequency. Shahruz, in 2006, proposed a methodology of designing a generator array by connecting cantilevers of different lengths which can increase the operation frequency range of the energy harvester. Further investigation is done by designing and simulating a cantilever array structure consisting of three cantilever beams having lengths of 2000, 2100 and 2200 μm, respectively. All the cantilevers have the same thickness of 2 μm and width 200 μm with a proof mass of length 200 μm, width 200 μm and thickness 10 μm. First cantilever having length 2200 μm resonates at a frequency of 681.061 Hz as shown in the Fig. 3.10. At frequency of 867.671 Hz, cantilever of length 2100 μm resonates, whereas at 1215 Hz cantilever of length 2000 μm resonates. The cantilever having length 2200 μm gives a maximum displacement as 5.7351 μm in Z-axis as shown in Fig. 3.9. The second and third cantilevers give zero displacements. Similarly, the second cantilever with length 2100 μm gives a displacement of 5.80 μm, and third cantilever having length 2000 μm gives displacement of 6.1 μm.

The above design of cantilever array has three elements of different lengths from 2000 to 2200 μm. The design gives an operation for frequency ranging from 681 to 1215 Hz. Therefore, a wide operation range is achieved using cantilever-array-type P-VEH. However, cantilever array design has multiple cantilevers which increase the cost of fabrication of the energy harvester. As in this case, only one cantilever is excited at a single frequency, the rest two cantilevers generate no or residual output. Due to this reason, the cantilever-array-type P-VEH has very low output-to-volume ratio. During device fabrication of cantilever-array-type P-VEH, releasing of multiple cantilevers having different dimensions is a challenge.

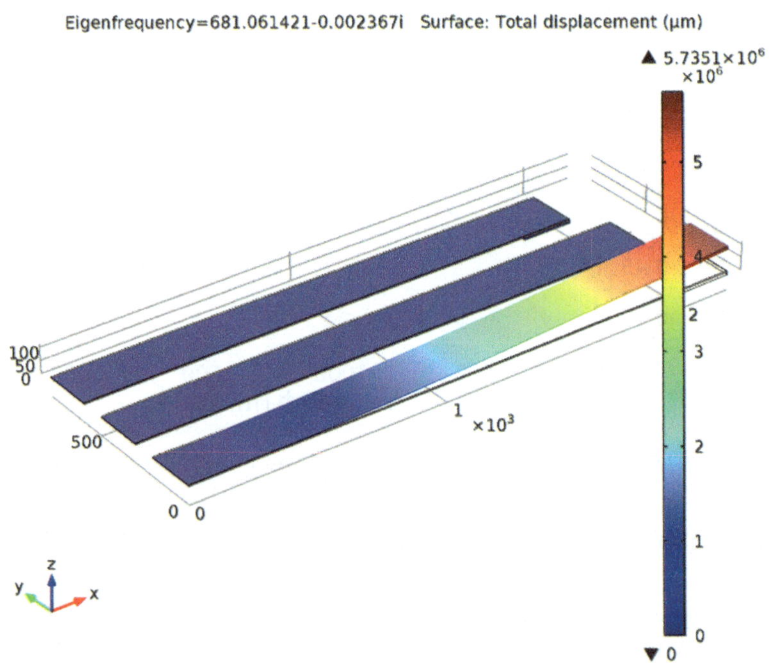

Fig. 3.9 Generator array of three cantilevers

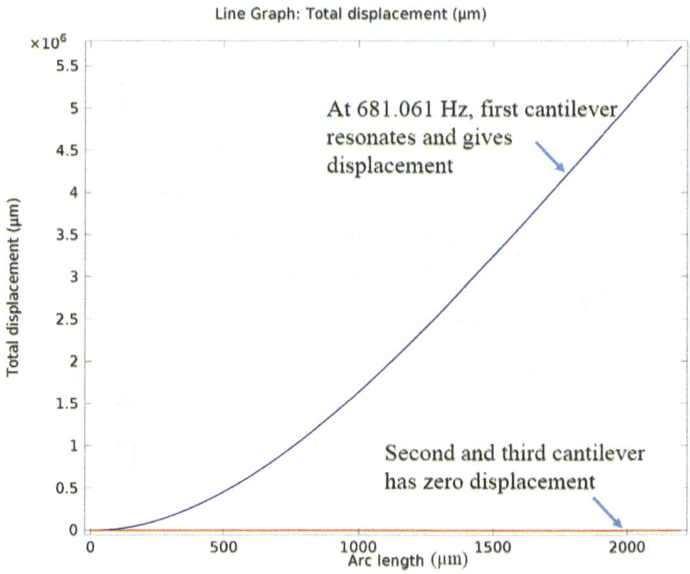

Fig. 3.10 Line graph of displacement of cantilever-array-type P-VEH

3.3 Clamped–Clamped or Guided Beam Structures

A clamped–clamped structure or guided beam structure is fixed from both ends and having seismic mass at the center. A clamped–clamped structure was first designed and fabricated by Marzencki et al. (2009) in 2009. In 2012, Wang et al. (2012) designed and fabricated a clamped–clamped structure with lower frequency operation of 694 Hz. Guided beam structures provide stable and reliable operation as they are fixed from both ends that enable to sustain higher stress and lesser displacement which reduces frequent collision from cavity. However, due to fixed nature of beams, guided beam structure has higher spring constant which results in higher resonance frequency. Guided two-beam device fabricated by Marzencki et al. (2009) has a resonance frequency of 1800 Hz and that developed by Wang et al. has 694 Hz (Wang et al. 2012). The resonance frequency for the guided two-beam device reported by Marzencki et al. (2009), Wang et al. (2012) is at higher end. Therefore, this motivated us to further investigate the two-beam guided cantilever structures for obtaining low frequency of operation by optimizing the design parameters.

A guided two-beam structure having dimensions as beam length 2500 width 2000 µm and thickness 20 µm, seismic mass having dimensions as 3500 × 3500 × 330 µm shown in Fig. 3.11 is simulated for obtaining the resonance frequency nature of displacement, stress and electric potential.

The nature of displacement of guided beam structures follows the following curve is given by Eq. 3.7 (Wang et al. 2012).

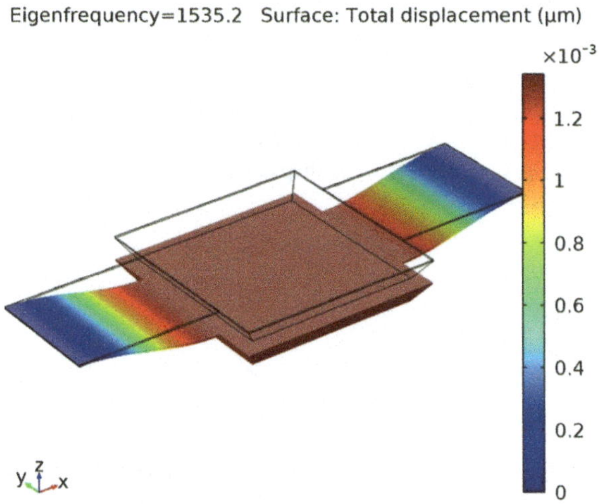

Fig. 3.11 Snapshot of eigenmode generated by the guided two-beam structure

$$Z(x) = \frac{d}{2}\left(1 - \cos\frac{\pi x}{l}\right) \tag{3.7}$$

The beam displaces from fixed end and also at the guided end as depicted from graph in Fig. 3.12. Guided beam structure is fixed at both ends and having seismic mass at the other end; therefore, collision from the encapsulation cavity of the guided beam structure is very less.

Von Mises stress of the guided beam structure is given in Fig. 3.13. There are

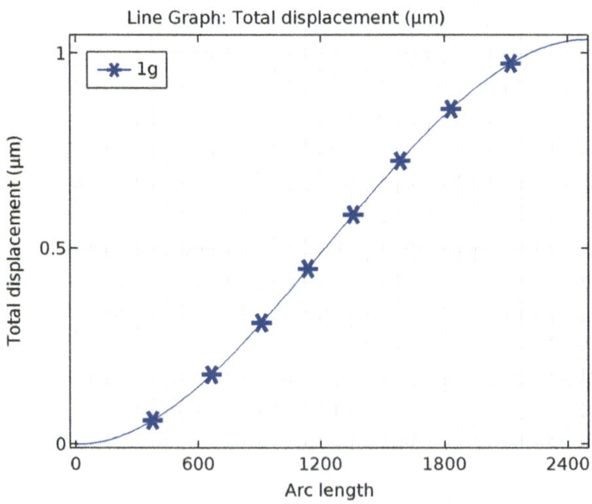

Fig. 3.12 Displacement versus arc length of guided beam-type structure

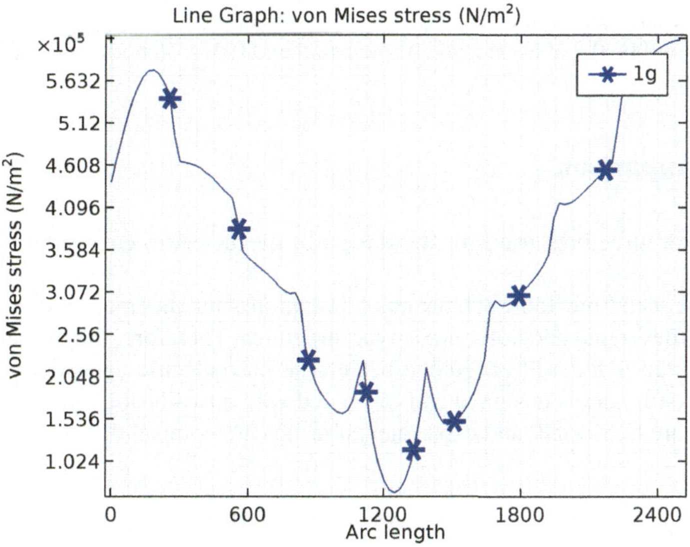

Fig. 3.13 Von Mises stress versus arc length of guided beam structure

two types of stress present tensile stress at fixed end and compressive stress at the guided end which are opposite in nature. Electric potential is generated due to tensile and compressive stress. Opposite charges are generated at the fixed end and guided end as seen from Fig. 3.14. At fixed end due to tensile stress, a maximum electric potential of 0.1 V is generated and at the guide end due to compressive stress 0.1 V of electric potential is generated. It can be seen that the individual potential generated

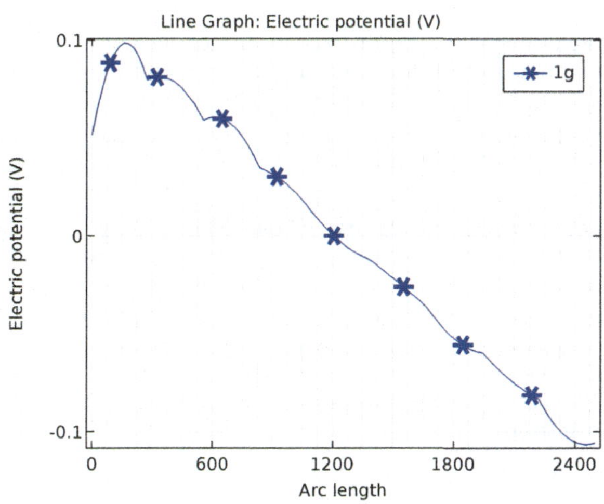

Fig. 3.14 Electric potential versus arc length of guided beam structure

at the fixed and guided end can be combined in series; therefore, the net potential generated will be 0.2 V at one side of the beam and 0.4 V on both sides of the beam.

3.4 Comparison

Three major structures used for the design of piezoelectric energy harvesters are compared.

It can be seen from Table 3.1 that guided beam structure has many advantages over single cantilever and generator array-type structures. Therefore, the further research work reported is focused on guided beam-type piezoelectric energy harvesters. A four-beam structure is designed and compared with a two-beam structure. Both the structures are fabricated, and experimental results are compared.

Table 3.1 Comparison of the three structures used as piezoelectric energy harvester

S. No.	Single beam	Generator array	Guided beam	Remarks
Displacement	8.5 μm	5.7351, 5.80, 6.1 μm	1 μm	Lower displacement gives better stability
Stress	6.5×10^6 N/m^2	6.5×10^6 N/m^2	5.6×10^5 N/m^2	Lower stress reduces risk of beam fracture
Electric potential	0.6 V	0.6 V	0.1 V and 0.4 V (net)	Higher electric potential better
Resonance frequency	764 Hz	681,867,1215 Hz	1535 Hz	Lower resonance frequency better
Residual stress	High because free from one end	High because free from one end	Low because fixed from both ends	Lower residual stress is required
Stability	Low (due to high displacement)	Low (due to high displacement)	High (due to less displacement)	Less displacement reduces collision from cavity
Reliable	Low (risk of beam fractures)	Low (risk of beam fractures)	Higher (fixed beams has lower risk)	Guided structure is more reliable
Ease of fabrication	Relatively complex due to fragile nature of free end cantilevers)	More complex due to multiple cantilevers release in fabrication	Less complex due to fixed beam structures	Guided beam structures can be designed and fabricated with less complexity

3.5 Effect of Shape of Seismic Mass on Potential Generated by Piezoelectric Energy Harvester

Seismic mass '*m*' is placed at the free end of the cantilever structure or at the center of the guided beam structure to reduce the resonance frequency of the energy harvester to match with ambient vibrations. Generally, the shape of the seismic mass designed and realized in MEMS are square in shape (Ishihara et al. 1999). The effect of shape of seismic mass on the potential generated by P-VEH has not been investigated. In this section, the effect of the shape of the seismic mass due to the change in the position of center of mass on output potential generated by P-VEH has been investigated. The geometry of the seismic mass will affect the change in the position of the center of mass of the system, resulting in a change in the potential generated. The center of mass for the device is analytically calculated by dividing the structure into the cantilever beam and the seismic mass. The center of mass for the two structures is individually calculated, and the vector sum is obtained to investigate the resultant effect of the center of mass of the seismic mass on the system. The geometry of the seismic mass is changed in such a manner that the width, depth and height of the seismic mass remain constant; therefore, the effect of shape can be investigated. The cantilever has a length: 2500 μm, width: 2000 μm and height 20 μm, respectively. The substrate layer is silicon which has a thickness of 20 μm. A piezoelectric layer of zinc oxide (ZnO) with a thickness of 2 μm is deposited on the silicon substrate. The seismic mass has dimensions from top as 3500 × 3500 μm, and height 330 μm, and slant ratio (r) which is a variable having a range between 0 and 1 that varies the bottom of the seismic mass from 0 to 3500 μm. The slant length of the shape of the seismic mass can be varied using variable r: when $r = 0$, it gives a maximum slant length which gives a triangular structure, when $r = 0.4$ it changes the shape to a pyramidal structure, and when $r = 1$ it forms a rectangular-shaped seismic mass, as shown in Figs. 3.15a–c, respectively. The center of mass of the energy harvester that has

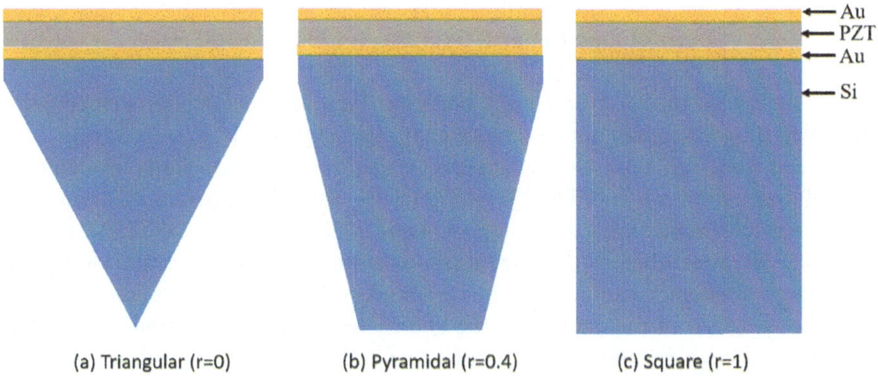

Fig. 3.15 **a** Shape of seismic mass (triangular) when the ratio $r = 0$, **b** the modified shape of seismic mass (pyramidal) at $r = 0.4$ and **c** the shape of seismic mass (rectangular) at $r = 1$

rectangular-shaped and triangular-shaped seismic mass is calculated and compared. The center of mass of the cantilever beam with dimensions length 2500 μm, width 2000 μm and height 20 μm will be at its centroid (1250, 1000, 10) because it is cuboid in shape. Similarly, the center of mass of the rectangular-shaped seismic mass that has dimensions length 3500 μm, width 3500 μm and height 330 μm (toward the negative *z*-axis) will be at its centroid (1750, 1750, −165).

The resultant center of mass of these two will be the center of mass of the energy harvester, which can be calculated using Eq. 3.8

$$\text{Center of mass}(\mu) = \frac{m_1 r_1 + m_2 r_2}{m_1 + m_2} \tag{3.8}$$

where m_1 and m_2 are the masses and r_1 and r_2 are the locations of the center of mass. The mass of the cantilever beam and seismic mass which can be calculated from the density of polysilicon is 2320 kg/m^3 and the volume of the cantilever beam is 10^8 μm^3 and that of the rectangular-shaped seismic mass is 4042.5×10^6 μm^3 which is calculated as 232×10^{-9} and 9378.6×10^{-9} kg, respectively. Calculating the center of mass using Eq. 3.8, the center of mass of the energy harvester with a rectangular-shaped seismic mass is obtained as:

$$\text{Center of mass}(\mu) = 1737.125\vec{x} + 1731.1\vec{y} - 160.75\vec{z} \tag{3.9}$$

The center of mass will be at a distance of *h*/4 from the base when the triangular-shaped seismic mass is considered, where *h* is the height of the triangle. The width and depth of the seismic mass remain unchanged; therefore, the center of mass for a triangular-shaped seismic mass will be (1750, 1750, −82.5), the volume of the triangular shape is 4042.5×10^6 μm^3, and the mass of the triangular-shaped seismic mass is calculated as 9378.6×10^{-9} kg. The center of mass for the cantilever beam remains the same. The resultant center of mass is calculated using Eq. 3.8, and the center of mass of the energy harvester with a triangular-shaped seismic mass is

$$\text{Center of mass}(\mu) = 1737.125\vec{x} + 1731.1\vec{y} - 80.50\vec{z} \tag{3.10}$$

The centers of mass for an energy harvester having a rectangular-shaped and a triangular-shaped seismic masses are given by Eqs. 3.9 and 3.10, respectively.

As seen from Table 3.2, the rectangular-shaped seismic mass shifts the center of mass of the cantilever by (1737.125x + 1731.1y − 160.75z), whereas the triangular-shaped seismic mass shifts the center of mass of the cantilever by (1737.125x + 1731.1y − 80.50z) which is (80.24z) above as compared with the center of mass of the rectangular-shaped cantilever. When a boundary load of 1 bar pressure is applied on the top surface of the cantilever, the energy harvester with a triangular-shaped seismic mass having center of mass shifted toward the top surface of the cantilever generates a higher potential as compared with the rectangular-shaped seismic mass. The motion of the cantilever structure is around its center of mass, and when center

Table 3.2 Comparison of different shape of seismic mass

Shapc	Center of mass of cantilever	Center of mass of seismic mass	Resultant center of mass
Rectangular shape	(1250, 1000, 10)	(1750, 1750, – 165)	1737.125, 1731.1, – 160.75
Triangular shape	(1250, 1000, 10)	(1750, 1750, – 82.5),	1737.125, 1731.1, – 80.50

of mass shifts upward or toward the point where force is applied, then a higher force is exerted on the center of mass resulting in higher generated potential.

3.6 Simulation of Different Geometries of Seismic Mass

COMSOL Multiphysics® simulation tool is used to verify the effect of the seismic mass shape of the energy harvester on the potential generated. A boundary load is applied on the top surface of the cantilever to obtain the results. Figure 3.16a represents the displacement along the surface of the cantilever, where displacement is maximum and the stress is minimum due to the motion of the cantilever at the free end. The red region depicts the maximum displacement and low value of the electric potential, the yellow region depicts the medium range of the potential generated, and the blue region at the fixed end of the cantilever exhibits the minimum deflection region having maximum stress and electric potential. This distribution of electric potential can be verified from the piezoelectric layer along which the output of the energy harvester is computed from (0 to 2500 μm) termed as the arc length, while the

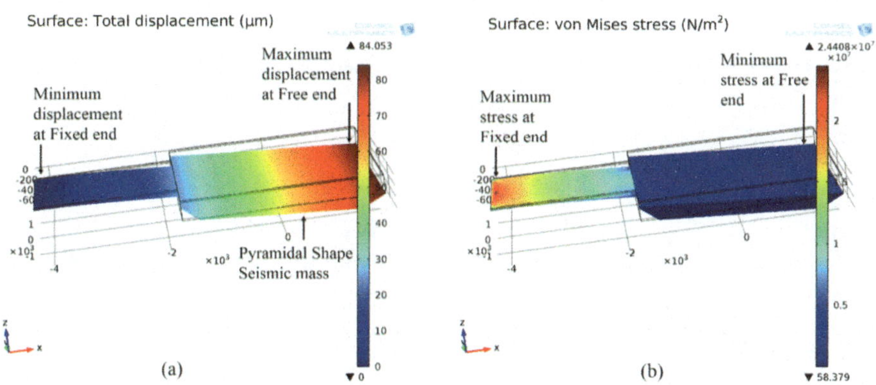

Fig. 3.16 **a** Displacement generated along the surface of the cantilever structure. **b** Distribution of von Mises stress along the surface of the cantilever structure

other side of the piezoelectric layer is grounded. Figure 3.16b shows the distribution of von Mises stress along the surface of the cantilever structure, where stress is maximum at the fixed end and minimum at the free end of the cantilever where value of stress is maximum in the region where there is minimum deflection of cantilever and decreases toward the free end of the cantilever. The potential is maximum at the fixed end of the cantilever because the stress is maximum at the fixed end of the cantilever.

Line graphs are correlated with the region of minimum deflection which produces maximum stress generating a maximum potential near the fixed end and discussed in the next section. The different shape of the seismic mass is investigated to study the effect of shape on the output parameters of the cantilever. Parameter '*r*', the ratio of the top and bottom length of the seismic mass, is varied from 1 to 0 with a step size of 0.2, which results in a variation from rectangular-shaped to triangular-shaped (with the pyramidal shape as the intermediate shape) of the seismic mass. Figure 3.17a represents the displacement of the cantilever along the arc length for different shapes (*r* varies from 0 to 1) of seismic mass. It can be seen that the displacement is minimum near the fixed end, and it increases nonlinearly toward the free end of the cantilever. Figure 3.17b gives an extended view of the displacement curve from 388 to 392 μm, and this region is considered because it is the region where the maximum potential is generated. For a triangular-shaped ($r = 0$) displacement at 392 μm is 0.519 μm and for a rectangular-shaped ($r = 1$) displacement at 392 μm it is 0.522 μm. Figure 3.18a gives the distribution of von Mises stress along the length of the cantilever, at the fixed end stress is maximum where the displacement is minimum. There is an abrupt variation near fixed end from arc length 0 to 10 μm, which again justifies the investigation results that at the fixed end the stress is maximum and displacement is almost negligible. Figure 3.18b gives the extended curve from 384 μm to 392 μm. The maximum stress is 1.771×10^7 N/m^2 for triangular shape ($r = 0$) and 1.773×10^7 N/m^2 for rectangular shape ($r = 1$).

Figure 3.19a gives the electric potential versus arc length curve from 0 to 2500 μm. Figure 3.19b is an extended view of the electric potential versus arc length curve from

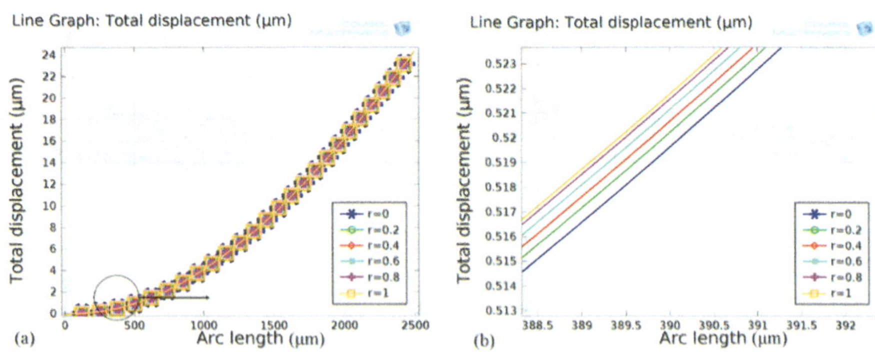

Fig. 3.17 **a** Displacement versus arc length graph for different values of *r*. **b** Displacement versus arc length graph extended curve from arc length 388–392 μm for different values of *r*

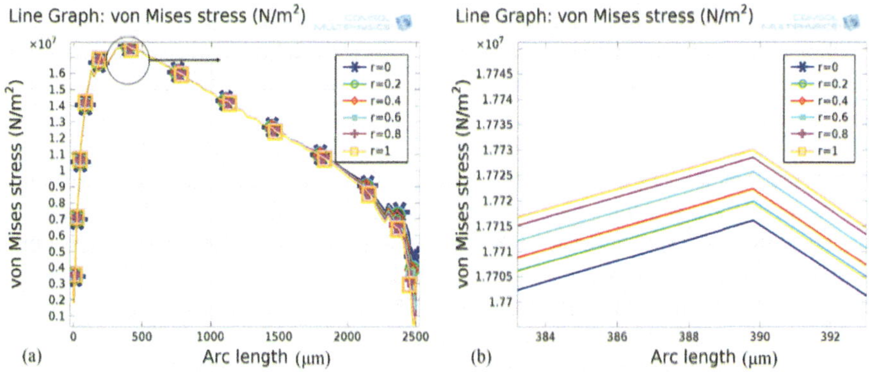

Fig. 3.18 **a** Von Mises stress versus arc length graph for different values of r. **b** Von Mises stress versus arc length graph extended curve from arc length 384–392 μm for different values of r

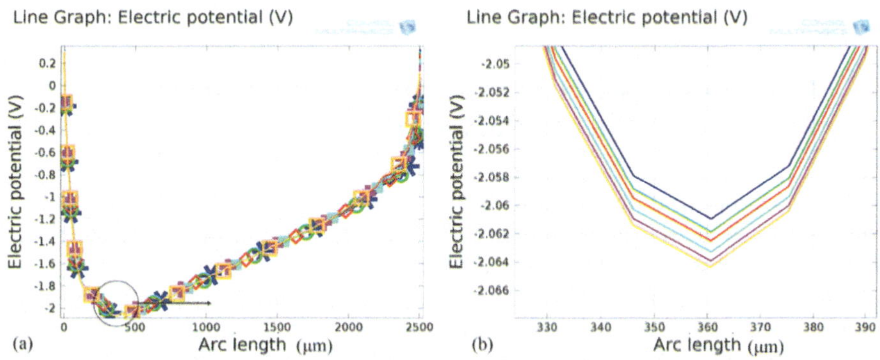

Fig. 3.19 **a** Electric potential versus arc length graph for different values of r. **b** Electric potential versus arc length graph extended curve from arc length 330 to 390 μm for different values of r

330 to 390 μm, where the cantilever having a triangular-shaped seismic mass generates a potential of −1.16 V, whereas a rectangular-shaped seismic mass generates −1.13 V (negative sign signifies the polarization of the piezoelectric layer with the base axis). It can be seen that the potential generated by the triangular-shaped seismic mass structure is higher than the potential generated by rectangular-shaped seismic mass by 0.03 V, which justifies the fact that there is considerable effect of the shape of seismic mass on the potential generated by the energy harvester.

3.7 Discussion

The effect of the shape of a seismic mass on the output parameters of a piezoelectric energy harvester is investigated, analyzed and compared. A triangular-shaped seismic mass generates a potential of 2.064 V, whereas a rectangular-shaped seismic mass cantilever structure generates a potential of 2.060 V, which clearly shows that there is a considerable increase in potential due to the variation in the shape of seismic mass.

The increased potential generated by the triangular-shaped seismic mass is due to the fact that the resultant center of mass of the geometry (cantilever structure and seismic mass section) gets shifted upward, i.e., toward the applied pressure, which results the exertion of a high force which leads to a higher potential being generated. However, to fabricate a triangular-shaped seismic mass (high slant ratio) requires a very high precision during bulk micromachining. Hence as an alternative, pyramidal-shaped seismic mass is used for the design and which can be realized using tetra-methyl ammonium hydroxide (TMAH) CMOS compatible 25 wt.% wet etching which gives an etch angle of 54.6 at <100> plane. In the next chapter, this pyramidal-shaped seismic mass is used to design guided two-beam and four-beam structure, and FEM results are obtained and compared.

3.8 Effect of Piezoelectric Material on Guided Beam Piezoelectric Energy Harvester

FEM simulation for different piezoelectric materials; ZnO, PZT-5H and AlN have been done, and the results are presented in Table 3.3 for guided four-beam P-VEH and in Table 3.4 for guided two-beam P-VEH.

It can be observed from the above FEM analysis that the ZnO piezoelectric material results in maximum potential generation for both guided two-beam and four-beam P-VEH. Therefore, ZnO material was selected during device fabrication.

Table 3.3 Comparison for different piezoelectric materials for guided four-beam P-VEH

Thickness = 2 μm	Displacement (μm)	Von Mises stress (N/m^2)	Potential (fixed end) (V)	Potential (guided end) (V)
ZnO	0.09889	97,801.476	**0.01779**	**−0.01819**
PZT-5H	0.1102	80,558.179	0.00231	−0.00233
AlN	0.0789	140,614.781	0.01147	−0.01175

Table 3.4 Comparison for different piezoelectric materials for guided two-beam P-VEH

Thickness = 2 μm	Displacement (μm)	Von Mises stress (N/m^2)	Potential (fixed end) (V)	Potential (guided end) (V)
ZnO	0.191	191,473.062	**0.03412**	**−0.03551**
PZT-5H	0.212	157,260.796	0.00442	−0.00455
AlN	0.151	273,237.3125	0.02187	−0.02272

References

Bao M (2005) Analysis and design principles of MEMS devices. Elsevier
Ishihara K, Yung CF, Ayón AA, Schmidt MA (1999) Inertial sensor technology using DRIE and wafer bonding with connecting capability. J Microelectromech Syst 8(4):403–408
Marzencki M, Defosseux M, Basrour S (2009) MEMS vibration energy harvesting devices with passive resonance frequency adaptation capability. J Microelectromech Syst 18(6):1444–1453
Shahruz SM (2006) Design of mechanical band-pass filters for energy scavenging. J Sound Vib 292(3–5):987–998
Wang Z, Matova S, Elfrink R, Jambunathan M, De Nooijer C, van Schaijk R, Vullers RJM (2012) A piezoelectric vibration harvester based on clamped-guided beams. In: 2012 IEEE 25th International Conference on Micro Electro Mechanical Systems (MEMS). IEEE, pp 1201–1204
Williams CB, Yates RB (1996) Analysis of a micro-electric generator for microsystems. Sens Actuat A Phys 52(1–3):8–11
Zhu D, Tudor MJ, Beeby SP (2009) Strategies for increasing the operating frequency range of vibration energy harvesters: a review. Meas Sci Technol 21(2):022001

Chapter 4
Design and FEM Simulation of Guided Beam Piezoelectric Energy Harvester

Abstract In this chapter, design, FEM analysis and comparison of guided two-beam and four-beam P-VEH have been discussed. Device parameters such as resonance frequency, beam displacement, stress and electric potential have been obtained and compared for both devices. A heavy pyramidal-shape seismic mass is connected at the center for higher electric potential and lower resonance frequency. A guided four-beam device can withstand higher stress and gives reliable operation as compared to a guided two-beam device. Design optimization for the split electrodes has been done for higher output potential and low-frequency operation. The optimized design for guided two-beam and four-beam device has been selected for fabrication in MEMS technology.

Keywords Four beam · ZnO · Split electrode · Seismic mass

Guided beam P-VEH fixed from four ends and having seismic mass at the center gives stable, reliable and improved response as compared to single-beam cantilever-type P-VEH. Seismic mass is supported with four beams which gives a stable operation and ability to sustain higher stress values due to high-amplitude vibrations as compared to single-beam cantilever structure (Marzencki et al. 2009). Four beams provide higher stiffness to the structure resulting in lower displacement which reduces frequent collision of the seismic mass with the encapsulation cavity avoiding beam fractures of the device. As the proof mass at the center is guided by four beams, this structure gives lower residual stress as compared to single-beam cantilever structure. Piezoelectric layer sandwiched between electrodes is deposited on four beams, which generates potential due to compressive and tensile stress. The potential generated is collected through the electrodes and can be used to power a small Li-ion battery or charge a super capacitor.

4.1 Design of Guided Four-Beam P-VEH

Guided four-beam P-VEH is designed to achieve lower resonance frequency and improved output response. Four beams are fixed form one end and connected with a

S. Saxena et al., *Design and Development of MEMS based Guided Beam Type Piezoelectric Energy Harvester*, Energy Systems in Electrical Engineering,
https://doi.org/10.1007/978-981-16-0606-9_4

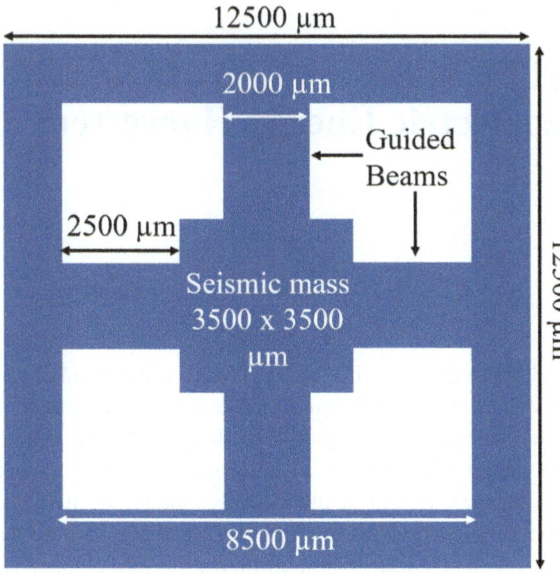

Fig. 4.1 Design structure for guided four-beam P-VEH

seismic mass at the center as shown in Fig. 4.1. Each beam has a length of 2500 μm and width of 2000 μm which gives a centrosymmetric structure. After several iterations, beam length of 2500 μm has been selected which gives a lower resonance frequency with reasonable device dimensions of 1.2 mm × 1.2 mm. Beam width is kept 2000 μm and thickness 20 μm because it gives support to the beams when suspended with a heavy seismic mass giving stability and reducing residual stress. Moreover, the beam thickness and width have to be selected at a lower side as it increases the resonance frequency of the device. A heavy seismic mass having dimensions of 3500 μm × 3500 μm and thickness 330 μm has been suspended through four beams as seen from Fig. 4.1. This reduces the resonance frequency of the device which is necessary to match with low-frequency ambient vibrations. Several iterations have been done to select the shape of seismic mass. Different dimensions 2500 μm × 2500 μm, 3500 μm × 3500 μm and 4500 μm × 4500 μm were designed, and FEM results were obtained. Higher output with higher stress handling capacity was shown when seismic mass dimensions were 3500 μm × 3500 μm and thickness 330 μm. Shape of the seismic mass is also an important factor as it affects the potential generated as discussed in the previous chapter. A pyramidal shape seismic mass provides higher generated potential than the square shape seismic mass.

This pyramidal shape of the seismic mass is realized through wet bulk micromachining using Tetra Methyl Ammonium Hydroxide (TMAH) etching with suitable corner compensation structures. Dimensions for the seismic mass at the top are

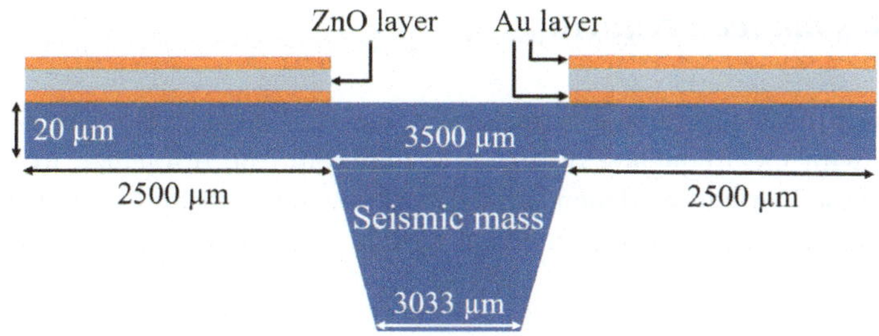

Fig. 4.2 Cross-sectional view of guided beam P-VEH

3500 μm × 3500 μm (Fig. 4.1), and from the bottom is 3033 × 3033 μm and height is 330 μm as shown in Fig. 4.2.

Figure 4.2 gives the cross-sectional view of the device giving the layer details. The base structure is of silicon giving beam thickness of 20 μm. Thermal oxide layer is grown for insulation between substrate and the bottom electrode. Gold (Au)/Aluminum (Al) metal is deposited and patterned over oxide layer to form the bottom electrodes. A piezoelectric layer of zinc oxide (ZnO) having c-axis orientation is deposited over bottom electrode. Gold (Au)/Aluminum (Al) metal is deposited over ZnO layer and patterned to form the bottom electrodes. Bottom and top electrodes (Au/Al) are sputtered over piezoelectric layer for charge collection. During device fabrication, a passivation layer of plasma enhanced chemical vapor deposition (PECVD) oxide of thickness 0.2 μm was deposited at the bottom and top surface of the ZnO layer to protect it from the environmental deterioration due to ambient oxygen and water vapor in the atmosphere.

A P-type silicon substrate (100) of 3 inches diameter having a thickness of 350 ± 25 μm is suitable for device fabrication. Zinc oxide material has been selected because of its compatibility with CMOS fabrication technology and higher piezoelectric coefficient. Ambient vibration displaces the guided beams which generate tensile and compressive stress in the beams resulting in the generation of charge along the piezoelectric layer due to piezoelectric effect.

4.2 FEM Analysis of Guided Four-Beam P-VEH

In this subsection, guided four-beam P-VEH is designed based on analytical equations and investigated using finite element method (FEM) simulator (COMSOL Multiphysics®). Parameters such as resonance frequency, total displacement, von Mises stress and output potential are obtained as a function of input acceleration.

4.2.1 Resonance Frequency

The first criterion to generate maximum output power is to match the device resonance frequency with the frequency of ambient vibration. The resonance frequency of cantilever-based structures depends on two parameters: spring constant (k) of the beam and seismic mass (m) suspended at the end or center as shown in Eq. (4.1).

$$f = \frac{1}{2\pi}\sqrt{\frac{k}{m}} \tag{4.1}$$

$$k = \frac{nEWt^3}{l^3} \text{ (for guided beam)} \tag{4.2}$$

Spring constant 'k' of the cantilever depends on four factors Young's modulus (E), width (W), thickness (t) and the length (l) of the cantilever shown in Eq. (4.2). 'n' is the integer that takes value of 2 for two-beam structure and 4 for four-beam structure. Resonance frequency can be reduced by choosing lower width and thickness and higher length of the beam. A heavy seismic mass (m) lowers the resonance frequency of the device necessary to tune with ambient vibrations.

Resonance frequency for the guided four-beam is obtained using FEM simulator for beam thickness from 5 to 20 μm. Length (l) and width (W) of the four guided beams are fixed at 2500 μm and 2000 μm, respectively. Seismic mass having dimensions 3500 × 3500 × 330 μm and weight 8.155×10^{-6} kg is attached at the center of the four beams to lower the resonance frequency. The snapshot of the device resonance frequency for beam thickness of 5 μm which comes at 335.96 Hz is shown in Fig. 4.3.

Analytical Eqs. (4.1) and (4.2) are used to calculate resonance frequency for a thickness of 5 μm of the silicon beam which comes out to be 337.72 Hz. As thickness of the silicon beam increases, the resonance frequency of the four-beam structure increases as shown in Eq. 4.2. When the beam thickness is increased to 10 μm, the resonance frequency increases to 698.87 Hz. Beam thickness is increased to a maximum value of 20 μm, the resonance frequency is maximum at 1631.4 Hz. Analytical calculation at a beam thickness of 20 μm gives resonance frequency of 1754.85 Hz. Figure 4.4 gives a comparison of the FEM and analytical results obtained by variation of beam thickness of the device. The results obtained using FEM and analytically are in close agreement with each other.

Resonance frequency obtained at different thickness of guided beam from 5 to 20 μm is in lower frequency region which is suitable for energy harvesting applications. Previous designs of the guided beam structures have resonance frequency of 1800 Hz (Marzencki et al. 2009) and 694 Hz (Wang et al. 2012). The results reported in Fig. 4.4 conclude that a guided four-beam structure is capable of operating at lower frequency range (below 500 Hz) so as to resonate with the frequency of the ambient vibrations. Therefore, the design suggested and reported is suitable for piezoelectric-type vibration energy harvesting applications which can be fabricated

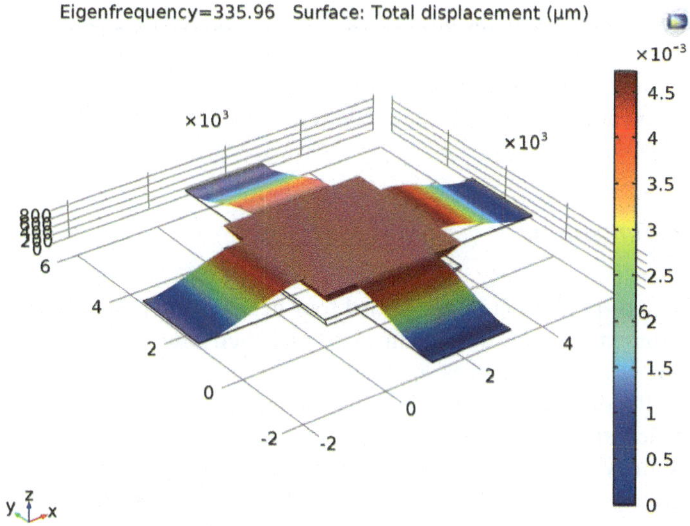

Fig. 4.3 Resonance frequency for four-beam guided structure at a beam thickness of 5 μm is 335.96 Hz

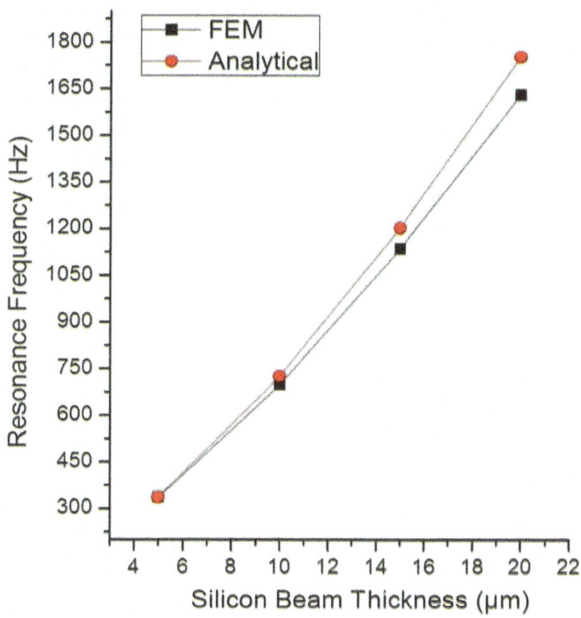

Fig. 4.4 Comparison of resonance frequency at a thickness from 5 to 20 μm of the silicon beam

using MEMS technology. In the proceeding sections, guided four-beam piezoelectric energy harvester is investigated with external vibrations. The acceleration is varied from a range from 1 to 5 g and parameters such as displacement, von Mises stress and electric potential are obtained and reported in the subsequent sections.

4.2.2 Displacement

In this section, displacement of the guided four-beam P-VEH due to external vibrations is discussed. Device displacement is observed after an input acceleration ranging from 1 to 5 g which is applied. The device gives the lowest resonance frequency or the dominant mode while displacing in the ±Z-direction as shown in Fig. 4.3. Arc length is defined as the length from fixed end of the beam to the guided end of the beam where seismic mass is connected. The arc position is selected as the interface between top surface of piezoelectric layer and the top electrode. Variation of the displacement along the arc length for the guided beam at an input acceleration ranging from 1 to 5 g is shown in Fig. 4.5. The graph shows that a displacement of 0.095 μm is obtained at an input acceleration of 1 g which increases with the input acceleration and attains a maximum value of 0.479 μm at an acceleration of 5 g. The graph in Fig. 4.5 shows that a maximum displacement is observed at the guided beam end which is attached with the seismic mass, and it gradually reduces while moving toward the fixed end of the beams. Near fixed end of the beam, stiffness is higher which results in lower displacement in that region and higher stress generation. To have the lowest resonant frequency, device operation should be in dominant mode

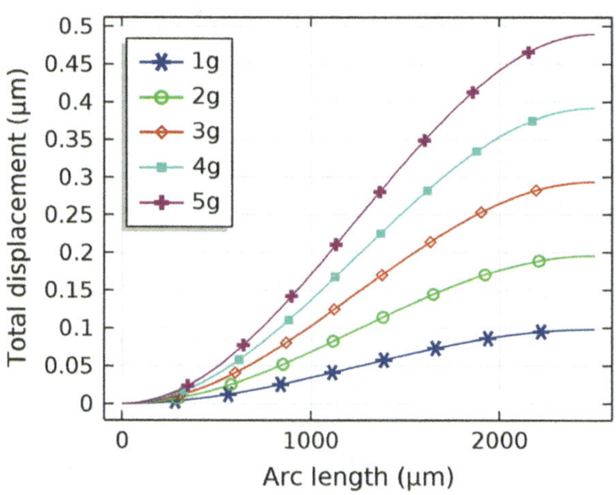

Fig. 4.5 Displacement versus arc length graph for the guided beam for input acceleration from 1 to 5 g. Displacement values of 0.095 μm and 0.48 μm are obtained at an input acceleration of 1 g and 5 g, respectively

to have resonance with low-frequency ambient vibrations; therefore, displacement curves for dominant mode are of significance.

Maximum displacement of the device helps in determining the depth of the cavity required for the device encapsulation. Larger displacement requires larger depth of the cavity otherwise due to collisions; fracture of beams will occur. A guided four-beam guided structure is fixed from four ends which results in lower displacement and leads to the stable operation of device; furthermore, the problem of residual stress is also not dominant in the reported structure as the beams have larger width and are fixed from four ends.

Displacement of the guided four-beam is also compared with the displacement of single-beam cantilever structure as shown in Fig. 4.6. One end of the cantilever is fixed, and seismic mass is placed at the other end. The dimensions of cantilever structure are length 2100 μm, width 200 μm and thickness 5 μm which is designed using FEM.

When an input acceleration of 1 g is applied, the single-beam cantilever gives a displacement of 0.14 μm which resonates at a frequency of 1611.9 Hz. However, the proposed guided beam structure gives a displacement of 0.095 μm at 1 g input acceleration which is less by 32.14% as compared to displacement obtained by single-beam cantilever structure.

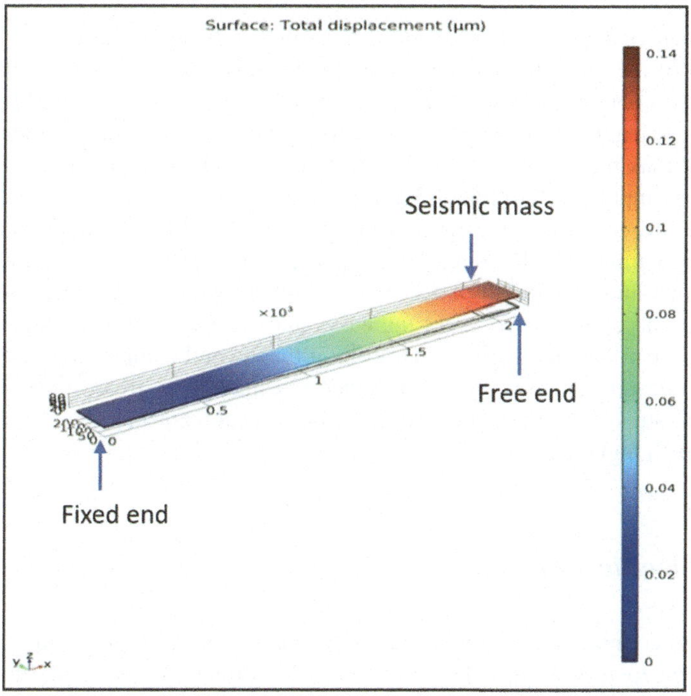

Fig. 4.6 Single-beam cantilever with seismic mass have a resonance frequency of 1611.9 Hz gives displacement of 0.14 μm at an input acceleration of 1 g

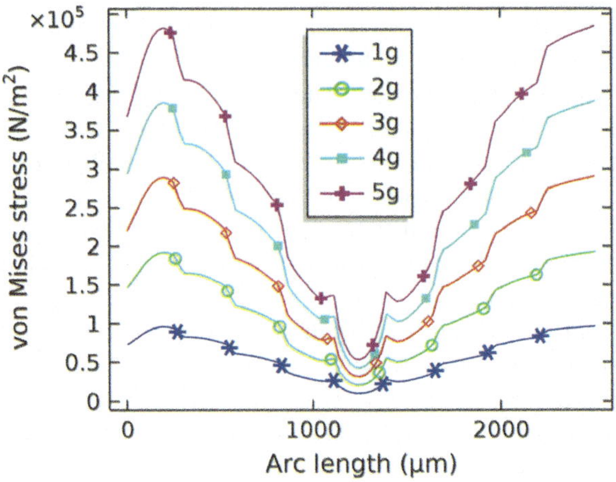

Fig. 4.7 von Mises stress versus arc length graph for the guided beam for input acceleration from 1 to 5 g. Maximum stress values of 9.4×10^4 N/m^2 and 4.7×10^5 N/m^2 are obtained at an input acceleration of 1 g and 5 g, respectively.

4.2.3 von Mises Stress

Displacement of the device in Z-direction generates stress along the X-axis, i.e., along the length of the guided beams. Fixed end generates tensile stress, and the guided end generates compressive stress along the beam. Figure 4.7 gives von Mises stress along the arc length graph for the guided four-beam for an input acceleration from 1 to 5 g. A maximum stress of 9.4×10^4 N/m^2 is observed at an input acceleration of 1 g. At an input acceleration of 5 g, maximum stress obtained is 4.7×10^5 N/m^2 across the piezoelectric layer (Fig. 4.7). Near the fixed end of the beam, tensile stress is maximum which decreases along the length of the beam and reaches to a minimum value at a length of 1250 µm. Further, as the length increases toward the guided end compressive stress increases and reaches to its maximum value at length 2500 µm which justifies generation of tensile and compressive stress shown in Fig. 4.7. Designed parameters of the device give maximum stress values well below the fracture stress for silicon; therefore, the design can withstand input acceleration of 5 g or even higher.

4.2.4 Electric Potential

Tensile and compressive stress in the guided beam results in generation of charge along the piezoelectric layer due to piezoelectric effect. Due to opposite nature of tensile and compressive stress, the generated charges are of opposite polarities along the length of the piezoelectric layer. Electric potential along the arc length of the beam

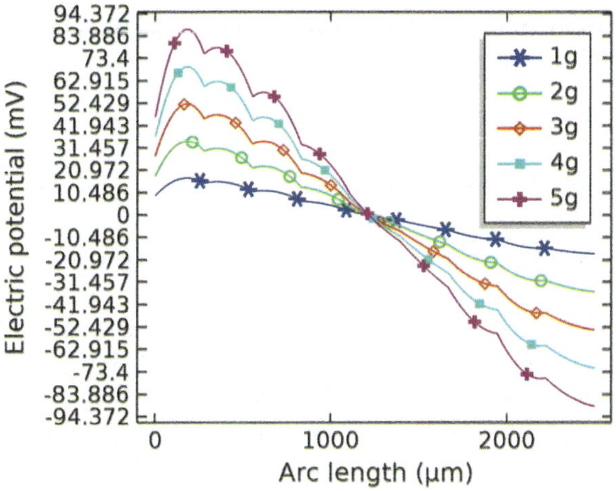

Fig. 4.8 Electric potential versus arc length graph for the guided beam for input acceleration from 1 to 5 g. Maximum electric potential of 17.84 mV and 89.21 mV is obtained at an input acceleration of 1 g and 5 g, respectively

for the guided four-beam structure for input acceleration from 1 to 5 g is shown in Fig. 4.8.

Maximum potential generated at an input acceleration of 1 g is 17.84 mV and at 5 g is 89.21 mV which can be seen from Fig. 4.8. Electric potential generated across the piezoelectric layer is maximum at the point of maximum stress (i.e., fixed end and guided beam end) which gradually reduces toward the minimum stress region (i.e., at the center of the beam). The graph justifies that electric potential is generated with opposite polarities due to tensile and compressive stress near fixed and guided beam end. The advantage of using guided beam structure is that two separate set of electrodes can be deployed to harvest this potential having opposite polarities, thus adding to the total potential harvested (Wang et al. 2012).

It can be observed that the potential generated due to tensile and compressive stress is quite similar in magnitude but with opposite polarities which are given in Table 4.1 for one beam. The same potential will be generated across the four beams

Table 4.1 Electric potential generated by one beam due to tensile and compressive stress at fixed end and guided end of four-beam guided cantilever structure

Input acceleration	Electric potential at fixed end (mV)	Electric potential at guided end (mV)
1 g	17.32	−17.84
2 g	34.68	−35.68
3 g	51.97	−53.52
4 g	69.29	−71.37
5 g	86.62	−89.21

of guided four beam structure which can be connected in series or in parallel to increase the net electric potential generated.

4.2.5 Selection of Seismic Mass

The device design is selected in such a way that it can be realized using available fabrication technology. Beam thickness of 20 μm is kept so that it can give stable and reliable operation and also withstands seismic mass. The maximum thickness of the seismic mass is fixed at 330 μm which makes 3-inch Si wafer having thickness 350 μm suitable for device fabrication. The length and the width of the seismic mass are optimized by selecting three values, i.e., 2500 × 2500 μm, 3500 × 3500 μm and 4500 × 4500 μm. Resonance frequency, stress and electric potential for the three variations are obtained using FEM. The dimensions of the seismic mass giving best results are further selected for device fabrication. The result of the comparison is presented in Table 4.2.

In case I, seismic mass of dimension 2500 × 2500 μm results in a resonance frequency of 2216.2 Hz and von Mises stress at 1 g is 0.8 × 10^5 N/m^2 and reaches to a maximum value of 4 × 10^5 N/m^2 at 5 g. Electric potential generated is 9.3 millivolts and 46.5 millivolts at 1 g and 5 g, respectively. Lower is the seismic mass '*m*' dimensions, higher is the resonance frequency as given by Eq. (4.1). Furthermore, it reduces the total displacement which further reduces the stress and potential generated.

In Case II, seismic mass with dimension of 3500 × 3500 μm gives a resonance frequency of 1631.4 Hz. von Mises stress at 1 g is 1.5 × 10^5 N/m^2 and reaches to a maximum value of 8 × 10^5 N/m^2 at 5 g. Electric potential generated is 17.32 mV and 86.62 mV at 1 g and 5 g, respectively. As this dimension of seismic mass gives considerable output with lower stress values hence chosen for designing the guided four-beam cantilever structure is suitable.

In case III, with seismic mass having dimension of 4500 × 4500 μm a resonance frequency of 1286.4 Hz is observed. von Mises stress at 1 g is 2 × 10^5 N/m^2 and reaches to a maximum value of 10 × 10^5 N/m^2 at 5 g. Electric potential generated is 28.64 millivolts and 143.2 mV at 1 g and 5 g, respectively. As the mass dimension is high, the resonance frequency generated is low but it results in higher stress values which may lead to fracture point of beams (Wang et al. 2012).

4.3 Comparison of Guided Two-Beam and Four-Beam Piezoelectric Energy Harvester

In this section, guided two-beam and four-beam P-VEHs are designed having same length, width, thickness and seismic mass. Parameters, such as resonance frequency,

Table 4.2 Displacement, von Mises stress and electric potential at different dimensions of seismic mass

	1 g	2 g	3 g	4 g	5 g
Case I 2500 x 2500 μm					
Displacement (μm)	0.050	0.100	0.140	0.200	0.250
von Mises stress (N/m^2, Maximum)	0.8×10^5	1.6×10^5	2.5×10^5	3×10^5	4×10^5
von Mises stress (Piezoelectric layer)	0.5×10^5	1.1×10^5	1.6×10^5	2.1×10^5	2.6×10^5
Electric potential (V)	0.0093	0.0186	0.0279	0.0372	0.0465
Case II 3500 x 3500 μm					
Displacement (μm)	0.095	0.190	0.286	0.382	0.479
von Mises stress (N/m^2, Maximum)	1.5×10^5	3×10^5	5×10^5	6×10^5	8×10^5
von Mises stress (Piezoelectric layer)	0.94×10^5	1.8×10^5	2.8×10^5	3.7×10^5	4.7×10^5
Electric potential (V)	0.01732	0.03468	0.05197	0.06929	0.08662
Case III 4500 x 4500 μm					
Displacement (μm)	0.140	0.300	0.450	0.600	0.700
von Mises stress (N/m^2, Maximum)	2×10^5	4.5×10^5	7×10^5	9×10^5	10×10^5
von Mises stress (Piezoelectric layer)	1.5×10^5	3×10^5	4.5×10^5	6×10^5	7.5×10^5
Electric potential (V)	0.02864	0.05729	0.08593	0.11458	0.14322

total displacement, von Mises stress, output potential, are obtained as a function of input acceleration and compared for the guided two-beam and four-beam structures, respectively.

4.3.1 Design, FEM and Comparison of Guided Two-Beam and Four-Beam Piezoelectric Energy Harvester

The guided two-beam and four-beam structures are fixed from both ends and have a pyramidal shaped seismic mass at the center as shown in Fig. 4.9a, b. The device has centrosymmetric nature with length of one beam is 2500 μm, width is 2000 μm and beam thickness of 25 μm. These beams are fixed from the end and supported with seismic mass at the center. A grid of 2000 μm is designed which act as fixed end for the beams and also provide area for contact pads for bottom and top electrodes.

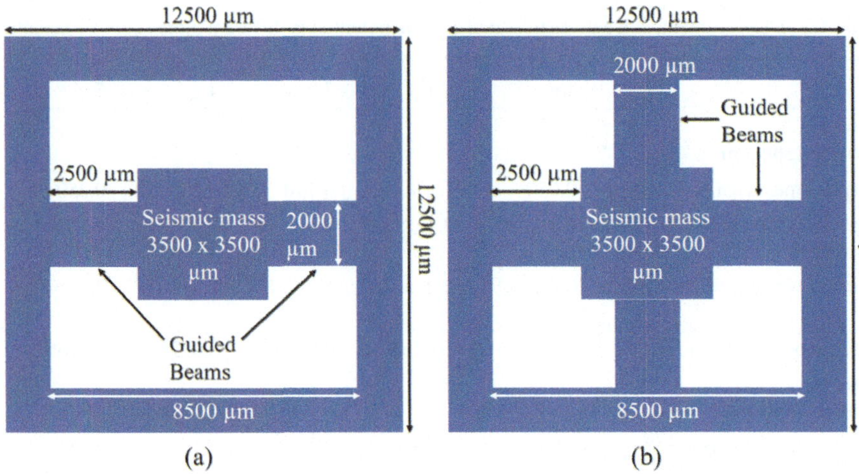

Fig. 4.9 Snapshot of design structure for **a** guided two-beam and **b** guided four-beam P-VEH

4.3.2 Resonance Frequency

As discussed, the first criterion for energy harvester to deliver optimum power is that the resonance frequency of the device should match with the frequency of ambient vibration. Energy harvester operating in resonance with the ambient vibration frequency gives maximum displacement which results in maximum output power.

This thickness of beam is selected as 25 μm for FEM eigenfrequency analysis and piezoelectric device study so that a realistic comparison from the experimental results can be done. Figure 4.10a, b gives the dominant mode resonance frequency obtained for the guided two-beam and four-beam structure. Two-beam structure gives

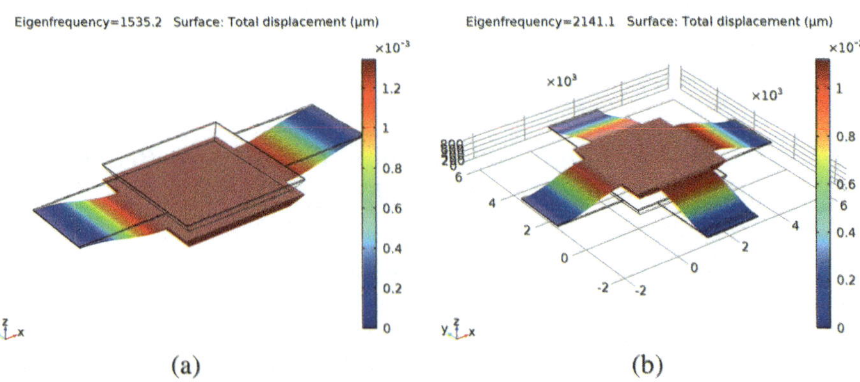

Fig. 4.10 **a** Snapshot of first mode at 1535.2 Hz for guided two-beam P-VEH. **b** Snapshot of first mode at 2141.1 Hz for guided four-beam P-VEH

maximum displacement in ±Z-direction at 1535.2 Hz, whereas the guided four-beam structure gives maximum displacement at 2141.1 Hz as shown in Fig. 4.10b. The resonance frequency calculated analytically using Eqs. (4.1) and (4.2) at beam thickness of 25 μm for the guided two-beam structure is 1622 Hz, whereas for the guided four-beam cantilever structure comes at 2294 Hz.

Second mode for the guided two-beam structure comes at 1798.9 Hz, whereas for guided four-beam structure second mode is at 2141.1 Hz. Third mode for the guided two-beam structure comes at 4618.8 Hz, whereas for guided four-beam structure third mode is at 35,203 Hz. The first mode of operation is the dominant mode having lowest resonance frequency at 1535.2 Hz for two-beam structure and 2141.1 Hz for four-beam structure.

The dominant mode of operation gives displacement in ±Z-direction which is the desired mode of operation for the energy harvesting device. Second mode is non-dominant mode along *y*-axis at 1798.9 and 2141.1 Hz which is not desired as it can severely affect the dominant mode device operation. Third mode is also non-dominant at 4618.8 Hz which is at higher frequency. It can be seen that, for the first mode the resonance frequency obtained from FEM (i.e., 1535.2 Hz) and calculated analytically (i.e., 1622 Hz) are in close approximation with each other.

4.3.3 Displacement

Ambient vibration frequency when match with the resonance frequency of the device then maximum displacement in ±Z-direction is obtained. In this section, displacement of the beam from the fixed end toward the guided end (where seismic mass is attached) is obtained for an input acceleration ranging from 1 to 5 g for the two structures. Figure 4.11a, b depicts the variation of displacement along the length

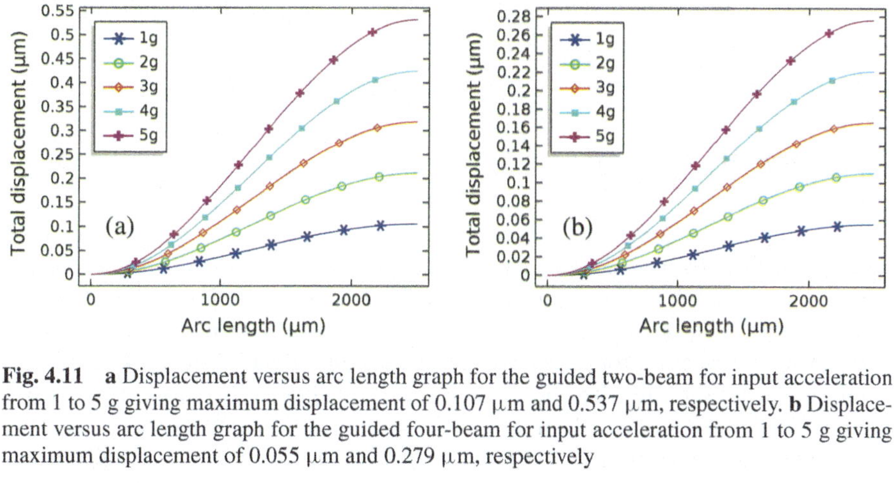

Fig. 4.11 **a** Displacement versus arc length graph for the guided two-beam for input acceleration from 1 to 5 g giving maximum displacement of 0.107 μm and 0.537 μm, respectively. **b** Displacement versus arc length graph for the guided four-beam for input acceleration from 1 to 5 g giving maximum displacement of 0.055 μm and 0.279 μm, respectively

Table 4.3 Comparison of maximum displacement at an input acceleration range from 1 to 5 g for guided two-beam and four-beam cantilever structure

Input acceleration	Maximum displacement for guided two-beam structure (μm)	Maximum displacement for guided four-beam structure (μm)
1 g	0.107	0.055
2 g	0.215	0.111
3 g	0.322	0.167
4 g	0.430	0.223
5 g	0.537	0.279

(arc length) of a beam from fixed end toward the guided end for the two-beam and four-beam structure, respectively. The graph justifies that the value of displacement is minimum at the fixed end which gradually increases toward the guided end attached with the seismic mass. The maximum displacement obtained at an input acceleration of 1 g is 0.107 μm, as the acceleration is increased displacement also increases which reaches to a maximum value of 0.537 μm at an acceleration of 5 g for guided two-beam cantilever structure as shown in Fig. 4.11a. Figure 4.11b gives maximum displacement obtained for guided four-beam cantilever structure at an input acceleration of 1 g is 0.055 μm, as the acceleration is increased displacement also increases which reaches to a maximum value of 0.279 μm at an acceleration of 5 g. Table 4.3 compares the maximum displacement for the guided two-beam and four-beam cantilever structure at an input acceleration range from 1 to 5 g. As four beams are attached with the seismic mass at the center provides higher stiffness (k) which results in lesser displacement (0.055 μm) as compared with guided two-beam cantilever structure which is approximately double as 0.107 μm at 1 g. As maximum displacement reduces, the probability of frequent collision with the encapsulation of the devices reduces.

Therefore, guided four-beam structure provides lesser displacement which reduces the risk of collision from the cavity typically at higher amplitudes, and therefore, gives stable operation as compared with guided two-beam structure.

4.3.4 von Mises Stress

Displacement in the guided beam generates stress along the X-axis or along the beam length. Tensile stress is generated at fixed end, and compressive stress is generated at the guided end of the beam. A maximum stress of 1.29×10^5 N/m^2 and 6.49×10^5 N/m^2 is generated at an input acceleration of 1 g and 5 g, respectively, across the length of the piezoelectric layer for guided two-beam structure as shown in Fig. 4.12a. At the fixed end, the tensile stress has maximum value which gradually reduces and reaches minimum value at the center (1250 μm) of the beam. As the length increases, from the center toward the guided end the compressive stress increases gradually and reaches

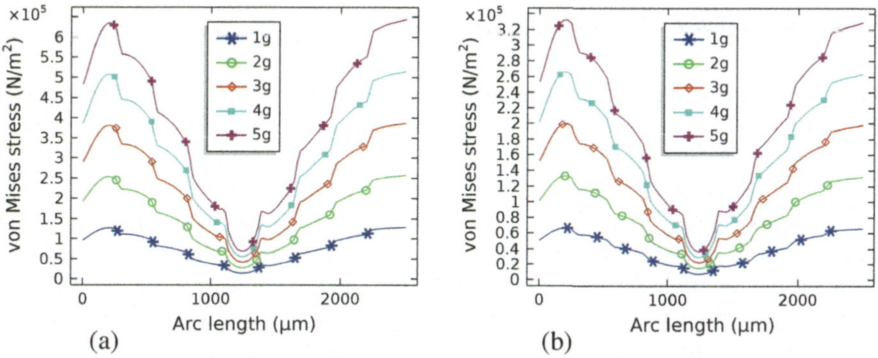

Fig. 4.12 **a** von Mises stress versus arc length graph for the guided two-beam for input acceleration ranging from 1 to 5 g giving maximum stress values of 1.29×10^5 N/m^2 and 6.49×10^5 N/m^2, respectively. **b** von Mises stress versus arc length graph for the guided four-beam for input acceleration ranging from 1 to 5 g giving maximum stress values of 0.67×10^5 N/m^2 and 3.3×10^5 N/m^2, respectively

maximum at guided end (2500 μm) which justifies the tensile and compressive stress shown in Fig. 4.12a.

Maximum stress of 0.67×10^5 N/m^2 and 3.3×10^5 N/m^2 is generated at an input acceleration of 1 g and 5 g, respectively, across the length of the piezoelectric layer for guided four-beam structure as shown in Fig. 4.12b. Table 4.4 gives the von Mises stress values for the guided two-beam and four-beam cantilever structure at an input acceleration range from 1 to 5 g. As four beams are attached with the seismic mass at the center, therefore, stress is equally distributed on the four beams generated due to deflection of the seismic mass. Therefore, each beam in guided four-beam structure exerts lesser stress, i.e., 0.67×10^5 N/m^2 (at 1 g) as compared to guided two-beam structure where each beam exerts a stress of 1.29×10^5 N/m^2. Reduced stress value will further reduce the risk of fracture of guided beams and therefore guided four-beam structure will provide a reliable operation as compared to guided two-beam structure.

Table 4.4 Comparison of von Mises stress at an input acceleration range from 1 to 5 g for guided two-beam and four-beam cantilever structure

Input acceleration	von Mises stress for two-beam structure (N/m^2)	von Mises stress for four-beam structure
1 g	1.29×10^5	0.67×10^5
2 g	2.59×10^5	1.3×10^5
3 g	3.89×10^5	2.01×10^5
4 g	5.19×10^5	2.6×10^5
5 g	6.49×10^5	3.3×10^5

4.3.5 Electric Potential

Stress across the length of the beam generates charge along the piezoelectric material due to piezoelectric effect. Charges of opposite nature are generated due to tensile and compressive stress at the fixed end and at the guided end of the beam. At the fixed end, a maximum potential of 23.62 mV is generated, whereas at the guided end, a maximum potential of −24.28 mV is generated at an input acceleration of 1 g for guided two-beam structure shown in Fig. 4.13a. Maximum potential generated at an input acceleration of 5 g is −121.38 mV for guided two-beam structure. Electric potential generated has maximum value at the fixed end due to maximum tensile stress which gradually reduces toward the center of the beam which is minimum stress region. Further, electric potential generated increases moving from center of the beam and is maximum at the guided end due to maximum compressive stress as shown in Fig. 4.13a.

Guided four-beam structure generates a maximum potential of −12.22 mV at 1 g and −61.12 mV at 5 g as shown in Fig. 4.13b. Table 4.5 compares the values of

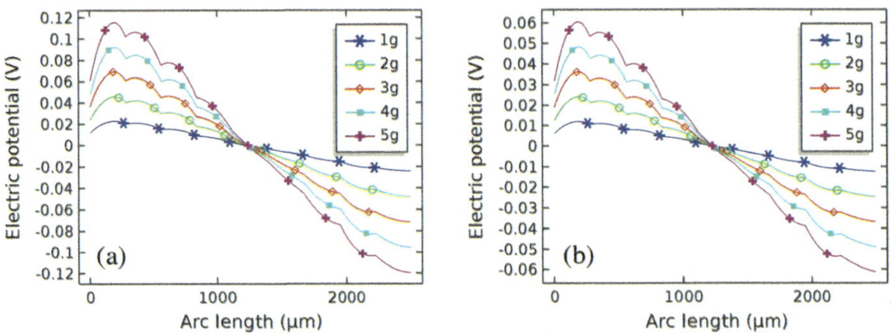

Fig. 4.13 **a** Electric potential versus arc length graph for the guided two-beam for input acceleration from 1 to 5 g giving maximum electric potential of 24.28 mV and 121.38 mV respectively. **b** Electric potential versus arc length graph for the guided four-beam for input acceleration from 1 to 5 g giving maximum electric potential of 12.22 mV and 61.12 mV, respectively

Table 4.5 Summarizes the electric potential generated by one beam of guided four-beam cantilever structure due to tensile and compressive stress at fixed end and guided end

Input acceleration	Electric potential at fixed end two-beam (mV)	Electric potential at guided end two-beam (mV)	Electric potential at fixed end four-beam (mV)	Electric potential at guided end four-beam (mV)
1 g	23.62	−24.28	12.05	−12.22
2 g	47.24	−48.55	24.11	−24.45
3 g	70.87	−72.83	36.16	−36.67
4 g	94.49	−97.11	48.21	−48.9
5 g	118.11	−121.38	60.27	−61.12

electric potential generated due to tensile and compressive stress at fixed end and guided end of one beam for guided two-beam and four-beam cantilever structure at an input acceleration ranging from 1 to 5 g, respectively. It can be seen from Table 4.5 that the electric potential generated across a beam for guided two-beam structure is 24.28 mV at 1 g which is twice the potential generated across a beam for guided four-beam structure which is 12.22 mV.

One beam of guided four-beam structure generates half the potential (12.22 mV) compared to guided two-beam structure but the net potential generated by the four beams would be 48.88 mV (4 × 12.22 mV) which is equal to the electric potential generated by two-beams of the guided two-beam cantilever structure.

The case discussed above is based on FEM results where the displacement is exactly double for guided two-beam structure to that of guided four-beam structure and could be considered as a special case. Experimental results discussed in next chapters show that the displacement for the two structures is not proportional and guided four-beam structure will generate higher potential.

From the results, it can be inferred that guided four-beam structure gives equal electric potential at smaller displacement and stress which improves the stability and reliability of device operation. Therefore, we conclude that a two-beam and four-beam structures of dimensions 2000 × 2500 μm with central pyramidal seismic mass of top dimensions 3500 × 3500 μm, bottom dimensions of 3033 × 3033 μm and height of 330 μm serve our purpose of generating potential in the range of millivolts under the ambient vibrations.

4.4 Design and Optimization of Split Electrodes for Guided Beam Piezoelectric Energy Harvester

As already discussed, two different tensile and compressive stress are present at the fixed and the guided end of the guided beam type piezoelectric energy harvester. Due to different stress, opposite polarities of charges are generated at fixed and guided end. Therefore, to harvest these charges set of split electrodes are required to collect charges. Marzencki et al. (2009) designed and fabricated a guided two-beam piezoelectric energy harvester by designing set of electrodes over the entire beam length. However, designing split electrodes is complex and this pattern needs to be optimized. Therefore, in this section we optimize the design of the split electrodes for the guided beam type piezoelectric energy harvester. Figure 4.14 shows variation of displacement and von Mises stress along the arc length of a beam which is investigated in detail to optimize the dimension and position for the split electrodes. Arc length is the length from fixed end of the beam (0 μm) to the guided end (2500 μm) of the beam where seismic mass is attached. The arc position selected is the interface of the top surface of piezoelectric layer and the top electrode where charge is generated and collected. Region A gives the beam displacement from the fixed end up to 675 μm as shown in Fig. 4.14. Tensile stress is generated along the beam length as shown in

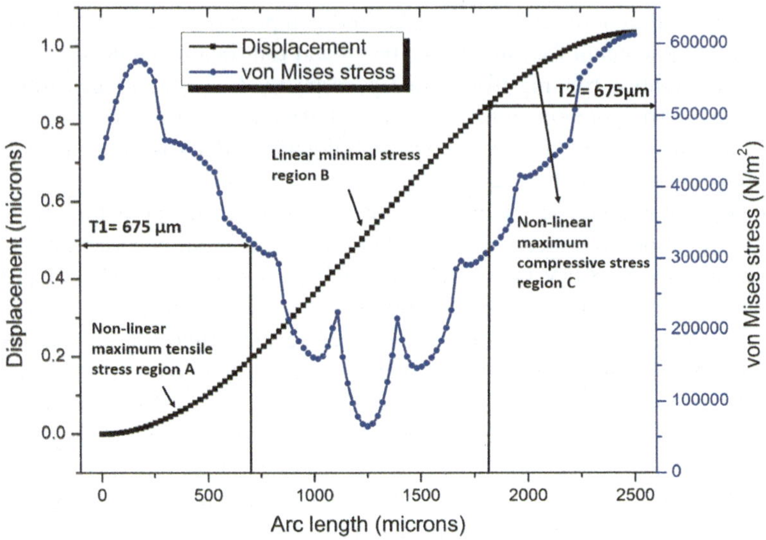

Fig. 4.14 Variation of displacement and von Mises stress along the beam length

region A which is governed by Eq. 4.3. Tensile stress reaches to a maximum value of 4.41234×10^5 N/m^2 at the fixed end which reduces while moving toward the free end.

$$f(x) = 9.567 \times 10^{-7} x^{1.867} \tag{4.3}$$

Split electrode $T1$ is placed from 0 to 675 μm so that the charge generated by the piezoelectric layer due to maximum stress can be harvested. Region B which is from 675 to 1825 μm of the beam length is a linear region governed by Eq. 4.4. Figure 4.14 shows that in region B both the tensile and compressive stress is minimum at 1250 μm which is the middle point of the beam.

$$f(x) = 0.001073x^{1.867} \tag{4.4}$$

Region C starts from 1825 to 2500 μm. Compressive stress is generated in the region C as shown in Fig. 4.14 which is governed by Eq. 4.5.

$$f(x) = -12.549 \times 10^5 x^{-4.126} \tag{4.5}$$

Split electrode $T2$ is placed between 1825 to 2500 μm at the guided end (where seismic mass is attached). This $T2$ electrode will collect charge generated by the piezoelectric layer due to maximum compressive stress. Exponentials from Eq. 4.3 for region A is 1.867 and for Eq. 4.5 for region C is −4.126. The difference sign and magnitude suggest that the two types of stresses are opposite and unequal in nature.

Due to higher displacement of seismic mass, compressive stress is higher as it is near to seismic mass. Compressive stress reaches to a maximum value of 6.12738 $\times 10^5$ N/m^2 at guided end which harvests a maximum potential of −0.10554 V. Tensile stress is dominant at the fixed end and reaches a maximum value of 4.41234 $\times 10^5$ N/m^2 harvesting a maximum potential of 0.098519 V. It is clear that, split electrodes $T1$ and $T2$ deployed in region A and B enable harvesting of maximum potential. Compressive stress at guided end is higher in comparison with tensile stress; therefore, higher potential is generated at the guided end. Split electrode designs used in the past literature such as discussed by Marzencki et al. (2009) have deployed set of electrodes on almost complete beam length; however, the approach presented here gives better results for harvesting ambient vibrations.

4.4.1 Effect of Length of Split Electrode on Resonance Frequency

In this section, effect of dimension of split electrodes on the resonance frequency of the guided two-beam piezoelectric energy harvester is discussed.

Three different electrode patterns have been considered as shown in Fig. 4.15. These patterns were designed on the guided two-beam structure and investigated using FEM. Electrode dimensions for Case I; length as 2500 μm, Case II; length as 675 μm and Case III; three electrode pattern used by Marzencki et al. (2009). Width is 2000 μm and thickness as 1 μm for all the three patterns, all other parameters are same as discussed in the design section. Table 4.6 compares the three patterns along with resonance frequency. Beam thickness of 5 μm gives resonance frequency of

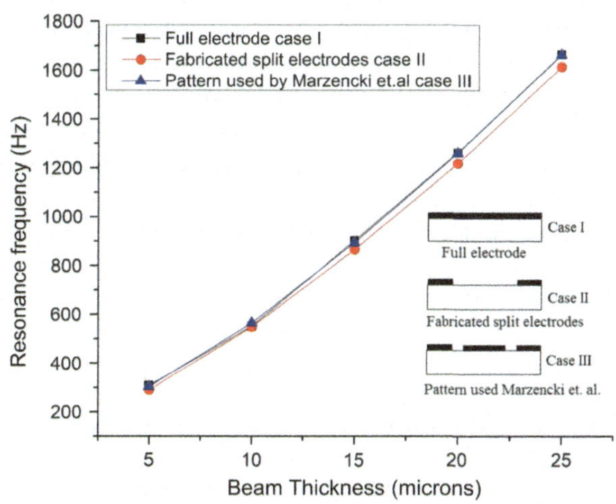

Fig. 4.15 Beam thickness versus resonance frequency at different electrode pattern

Table 4.6 Resonance frequency comparison for different electrode patterns

S. No.	$t = 5$ μm (Hz)	$t = 25$ μm (Hz)	Remarks
Case I	308	1664	4.2% less than case III
Case II	290	1613	
Case III	303	1663	

308 Hz for case I, 290 Hz for Case II and 303 Hz for Case III. Resonance frequency in case II is lower by 18 Hz or 5.8% as compared with case I and 13 Hz or 4.2% as compared with case III which can be seen from Fig. 4.15. Beam thickness of 25 μm gives resonance frequency of 1664 Hz for Case I, 1613 Hz for Case II and 1663 Hz for Case III. Resonance frequency for Case II is lower by 51 Hz or 3% by case I and 50 Hz or 3% by case III as shown in Fig. 4.15. Thus, Case II electrode pattern exhibits the least resonance frequency among the three cases. Therefore, Case II electrode pattern gives improved results, and hence, this pattern is selected during mask layout design for device fabrication.

4.4.2 *Effect of Length of Split Electrode on Change in Capacitance*

As the piezoelectric layer is sandwiched between the top and bottom electrodes; therefore, it forms a capacitor. Displacement of the beam from the seismic mass end will result in the change in capacitance which is proportional to the charge generated. The effect of variation of the length of electrodes on the beam displacement and capacitance of the guided two-beam P-VEH is investigated. Electrode length is varied from 675 to 2500 μm, and displacement of the guided two-beam P-VEH is obtained using FEM and presented in Fig. 4.16. At 1 g acceleration, beam displacement of 3.11422 μm is observed for electrode length of 675. Displacement gradually reduces with increase in electrode length, and minimum displacement of 2.51441 for electrode length of 2500 μm is observed. Thus, it can be concluded that as the length of the electrode is increased, the beam displacement reduces. The maximum variation of displacement observed is 0.59981 μm or 19.26%.

The change in capacitance for the piezoelectric energy harvester is given by Eq. (4.6)

$$\Delta C = \frac{xC}{d} \tag{4.6}$$

where ΔC is the change in capacitance, x is the displacement of the beams, C is the static capacitance (which is 153 pF experimentally obtained from capacitance–voltage (CV) test), d is distance between the electrodes which is 2.5 μm (ZnO layer). Figure 4.16 gives the change in capacitance (ΔC) due to variation in length of electrodes from 675 to 2500 μm. Electrode length of 675 μm gives maximum change

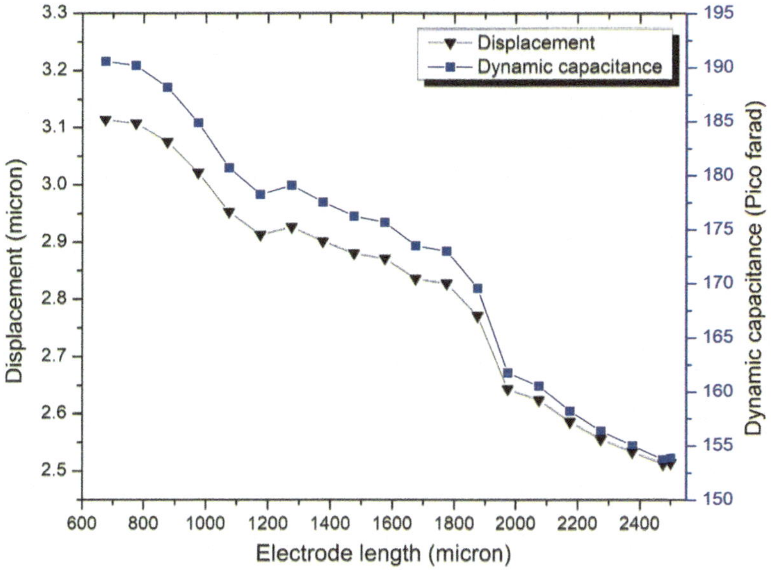

Fig. 4.16 Displacement and capacitance due to variation in electrode length

in capacitance, i.e., 190.59026 pF, which reduces as length of electrode increases, and reaches a minimum value of 153.88189 pF at an electrode length of 2500 μm. It can be stated that for an electrode length of 675 μm, change in capacitance is 36.70837 pF higher as compared to change in capacitance for electrode length 2500 μm.

Therefore, split electrodes of length 675 μm give optimum displacement and change in capacitance resulting in higher electric potential.

The above investigation clearly states that the optimized pattern of split electrodes gives better results by reducing the resonance frequency and by giving higher electric potential. Therefore, this optimized pattern is selected for device fabrication. The individual sensitivity of the split electrodes is also experimentally obtained during vibration bench testing which is discussed in detail in Chap. 6.

References

Marzencki M, Defosseux M, Basrour S (2009) MEMS vibration energy harvesting devices with passive resonance frequency adaptation capability. J Microelectromech Syst 18(6):1444–1453

Wang Z, Matova S, Elfrink R, Jambunathan M, De Nooijer C, van Schaijk R, Vullers RJM (2012) A piezoelectric vibration harvester based on clamped-guided beams. In: 2012 IEEE 25th International Conference on Micro Electro Mechanical Systems (MEMS). IEEE, pp 1201–1204, Jan 2012

Chapter 5
Fabrication of Guided Beam Piezoelectric Energy Harvester

Abstract This chapter reports fabrication of guided two-beam and four-beam P-VEH in MEMS technology. A five-level mask process has been designed for device fabrication. TMAH CMOS compatible 25 wt.% wet etching has been used for the realization of pyramidal-shape seismic mass and DRIE for releasing the two-beam and four-beam structures. Corner compensation is used to obtain perfect edges at the bottom vertex for pyramidal-shaped seismic mass. A 2.5-μm-thick ZnO layer is sandwiched between the bottom and top electrodes for the generation and collection of electric potential. Beam thinning using DRIE is done from the backside of the silicon wafer to reduce the beam thickness resulting in a significant reduction in the resonance frequency of the devices. A special kind of PCB was designed having gold-plated pads suitable for wire bonding, and also, a special kind of PCB attachment was designed with a built-in connector and attached to the PCB so that the coaxial cable can be connected to the analyzer.

Keywords TMAH · DRIE · ZnO · Bulkmicromachining · Seismic mass · Lithography

5.1 Introduction

Fabrication technology is one of the most significant steps in the realization of a device. It is the technology which puts forward several kinds of limitations in terms of materials used in the device, their processibility, growth, deposition and patterning, realization of mechanical structures, mechanical and physical properties of the material layers, fabrication of electrical and electronic components. A great care is required to develop and to characterize each unit process leading to the technology development for the fabrication of a specific device. The successful technology development is said to be complete when it delivers repeatable and reliable results. Fabrication of MEMS vibration energy harvester, apart from the usual unit processes, which are used in the fabrication of general MEMS devices, requires stringent micromachining techniques (Liu 2012) for both the devices. However, their fabrication is compatible with our technology and available processing facilities.

S. Saxena et al., *Design and Development of MEMS based Guided Beam Type Piezoelectric Energy Harvester*, Energy Systems in Electrical Engineering,
https://doi.org/10.1007/978-981-16-0606-9_5

In both of these devices, a combination of wet and dry bulk micromachining techniques has been utilized. Elaborate experimentation has been done on the development of micromachining processes based on both wet and dry etching techniques in which aqueous TMAH etching for wet bulk micromachining, reactive ion etching (RIE) and deep reactive ion etching (DRIE) have been used for dry bulk micromachining processes.

5.1.1 Fabrication Methodology

The fabrication methodology has to be very well planned as mechanical as well as electrical and piezoelectronic subsystems have to be fabricated simultaneously. For the current devices, the fabrication technology is somewhat different and hence need a special care in their development.

In this chapter, fabrication of guided two-beam and four-beam piezoelectric energy harvester is presented in detail.

5.2 Design of Mask Layouts

Each fabrication process starts with a set of masks which are used for the fabrication of each material layer or structure in the device. After fixing the dimensions of different structures and layers in the device through design and simulation in Chap. 4, the device has to be fabricated using MEMS fabrication technology. We need different masks for the fabrication of each layer or a structure. A five-level mask process was designed to realize the device. All these masks are summarized in Table 5.1.

The combined view of all the five-level masks for two-beam and four-beam devices designed using L-edit is shown in Fig. 5.1a, b. Mask #1 is for wet TMAH etching for pyramidal-shaped seismic mass. Mask #2 is for bottom electrode

Table 5.1 Mask details—Device: MEMS-VEH: Die size: 12.5 × 12.5 mm

Sr. no	Mask ID	Layer name	Mask polarity	GDS II No	Function of the mask
1	VEH-1	Active	Bright field	GDS_02_DT_00	TMAH etching
2	VEH-2	Metal 1	Bright field	GDS_10_DT_00	Bottom electrode patterning
3	VEH-3	Metal 4	Bright field	GDS_16_DT_00	Piezoelectric layer patterning
4	VEH-4	Poly	Bright field	GDS_04_DT_00	Top electrode patterning
5	VEH-5	Metal 2	Dark field	GDS_12_DT_00	DRIE for releasing the structure

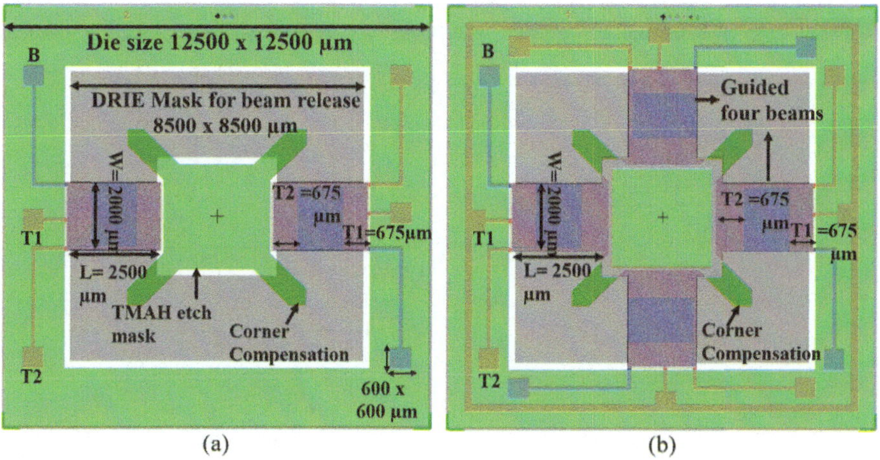

Fig. 5.1 5 Level mask layout for **a** guided two-beam device and **b** guided four-beam device

patterning. Mask #3 is for ZnO patterning. Mask #4 is for top electrode patterning having dimensions as 675 µm for each split electrode T1 and T2. Mask #5 is for release of guided beams using DRIE from the front side. Grid of 2000 µm is designed to define the fixed end of the beams and to define the electrode pads of 600 × 600 µm on both sides as shown in Fig. 5.1a. These masks are now described in detail.

5.2.1 Mask #1:TMAH Etch Mask for Pyramidal-Shaped Seismic Mass

As already stated, the pyramidal-shaped seismic mass in our design is realized using TMAH CMOS compatible 25 wt.% wet etching because it gives an etch angle of 54.74° for <100> plane. Mask #1 is designed for TMAH etching to realize pyramidal-shaped seismic mass. Figure 5.2a gives the layout of TMAH opening mask from the back side. The designed mask is a bright field mask, the bottom opening area for the TMAH mask is calculated from Eq. 5.1, where b1 is the bottom side opening area, b0 is from front side area and t_d is etch depth.

$$b1 = b0 + \sqrt{2}t_d \tag{5.1}$$

For an etch depth (t_d) of 330 µm and b0 as 8500 × 8500 µm, bottom opening b1 is calculated 8966.690 × 8966.690 µm, where b0 is the dimension from front side, i.e., 8500 × 8500 µm. Lateral etching during TMAH process must be taken into account which gives the bottom opening of 8916 × 8916 µm. Similarly, the seismic mass dimension from bottom is also calculated using Eq. 5.1 and after lateral etching

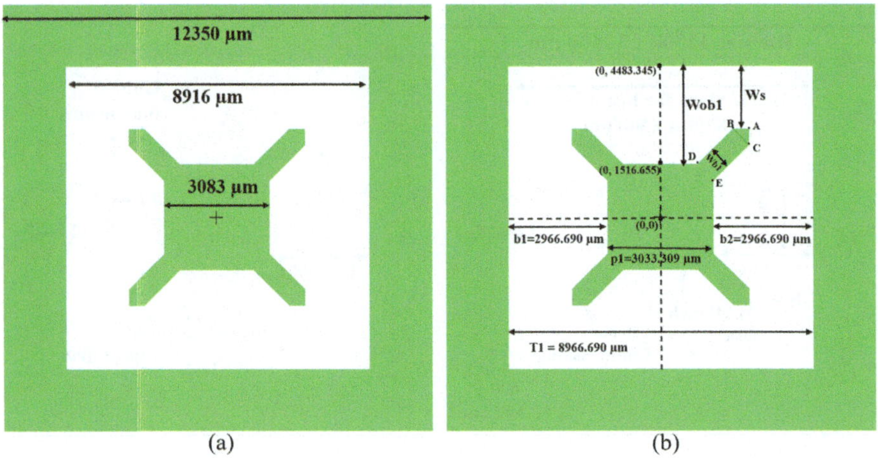

Fig. 5.2 **a** TMAH opening mask #1 for pyramidal-shaped seismic mass. **b** Corner compensation design to obtain perfect edges of seismic mass

compensation gives dimensions of seismic mass as 3083 × 3083 μm from the bottom side as seen from Fig. 5.2a.

5.2.1.1 Corner Compensation

As we know that convex corners of a structure become round or rather de-shaped because of the presence of high index planes at the corners in wet bulk micromachining, we need to mask these corners with structures which are known corner compensating structures. Therefore, corner compensation is used to obtain perfect edges at the bottom vertex for pyramidal-shaped seismic mass. Thin bar corner compensation is used for the energy harvester device (mukhiya et al. 2006). From Fig. 5.2b gives dimensions of Wob1 as 2966.690 μm, Wb1 the width of the bar is 660 μm (2td). Ws is calculated from Eq. 5.2 having dimension 1926.127 μm, where ratio (R(311)/R(100)) is 2.23.

$$W_{ob1} = W_s + 1.414\frac{R(311)}{R(100)}t_d \tag{5.2}$$

From Ws, coordinates for A point are obtained as (2557.217, 2557.217). Point B is calculated as (2090.526, 2557.217) (cos45° × 660 = 466.690)—similarly, point C, (2557.217, 2090.526), point D, (1049.964, 1516.654) and point E as (1516.654, 1049.964) respectively. Length of the bar (BD) is obtained from the coordinates and is calculated as 1471.577 μm.

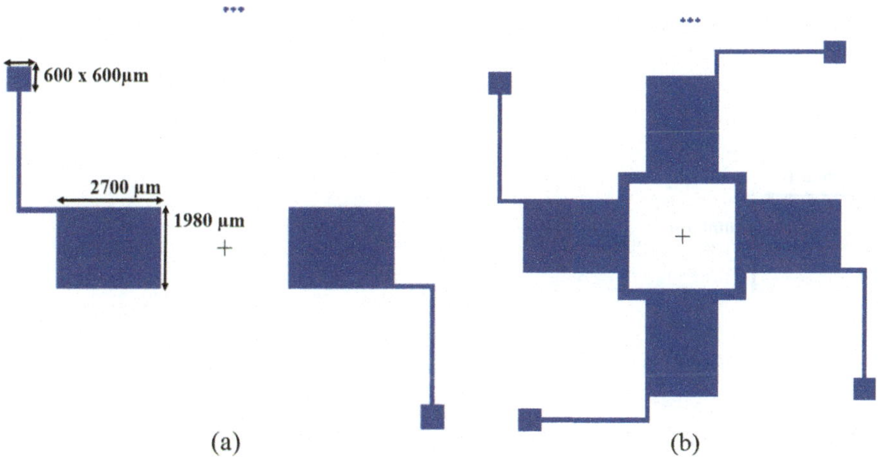

Fig. 5.3 Mask #2 for bottom electrode patterning for **a** two-beam device and **b** four-beam device

5.2.2 *Mask #2: Bottom Electrode Patterning*

Mask #2 is a bright field mask designed for bottom electrode patterning. The electrodes are placed on the guided two-beam and four-beam device. The dimensions of the electrodes are 2700 × 1980 μm, and the pads for connections are 600 × 600 μm as shown in Fig. 5.3a, b.

5.2.3 *Mask #3: Piezoelectric Layer Patterning*

Mask #3 is a bright field mask designed for piezoelectric layer patterning. The piezoelectric film is patterned on the beams having dimensions 2800 × 2000 μm as shown in Fig. 5.4a, b.

5.2.4 *Mask #4: Top Electrode Patterning*

Mask #4 is a bright field mask designed for top electrode patterning. The split electrodes are placed on the guided two-beam and four-beam device. Electrode T1 is at the fixed end of the beam, and electrode T2 is at the guided end of the beam. The dimensions of both electrodes T1 and T2 are 675 μm as shown in Fig. 5.5a, b.

Fig. 5.4 Mask #3 for piezoelectric layer patterning **a** two-beam device and **b** four-beam device

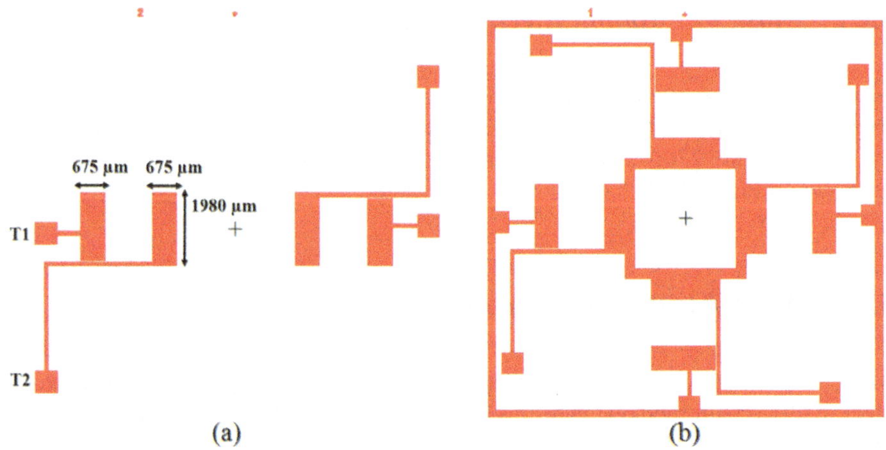

Fig. 5.5 Mask #4 for top electrode patterning a two-beam device and b four-beam device

5.2.5 *Mask #5: DRIE Etch*

Mask #5 is a dark field mask designed for beam release using DRIE from front side. The two and four beams are realized by etching the silicon using DRIE from front side using mask patterns as shown in Fig. 5.6a, b.

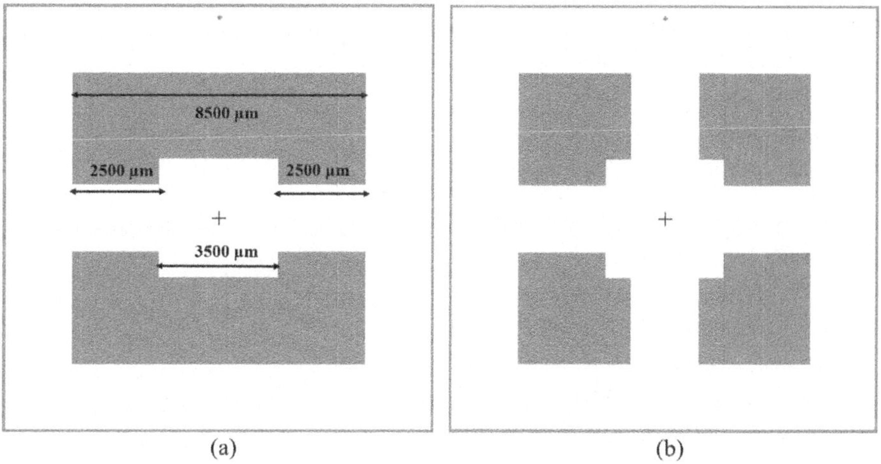

Fig. 5.6 Mask #5 DRIE etching **a** two-beam device and **b** four-beam device

5.3 Fabrication of Guided Beam Piezoelectric Energy Harvester

Guided two-beam and four-beam harvester devices are fabricated using MEMS fabrication processes. Bulk micromachining is done using tetra-methyl ammonium hydroxide (TMAH) CMOS compatible 25 wt.% wet etching to realize pyramidal-shaped seismic mass, and deep reactive ion etching (DRIE) is done from front side to release two-beam and four-beam structures in the device.

5.3.1 Device Fabrication Flow

The major process steps for the fabrication of the energy harvester device are shown in Table 5.2, whereas the fabrication process flow is illustrated in Fig. 5.7.

5.3.2 Device Fabrication (Batch 1)

Device fabrication was started with a batch of six number of P-type silicon single crystal <100> double-side polished (DSP) wafers of diameter 3 inches and thickness 350 ± 25 μm and having resistivity 1–10 Ω cm. These specifications of the wafers were selected on the basis of their suitability to realize the designed dimensions of pyramidal-shaped seismic mass using wet TMAH etching (mukhiya et al. 2006). The device fabrication is carried out as per the process flow steps in Table 5.2. Thorough

Table 5.2 Fabrication steps

1.	Thorough cleaning of wafers: degreasing and chemical cleaning
2.	Wet oxidation: 0.6 μm thick
3.	Deposition of LPCVD nitride 0.2 μm
4.	Back side patterning **(Lithography M#1)** and Oxide + nitride etching: definition of proof mass
5.	Wet bulk micromachining (WBMM) in TMAH: formation of proof mass (mukhiya et al. 2006)
6.	Wet etching of LPCVD nitride and oxide
7.	Dry oxidation: quality oxide for electrical isolation
8.	RF sputtering: Cr/Au for bottom electrode
9.	Patterning **(Lithography M#2)** and etching of Cr/Au: Bottom electrode formation
10.	Deposition of ZnO piezoelectric layer by reactive sputtering of Zn
11.	Patterning of ZnO **(Lithography M#3)** (Prasad et al. 2013)
12.	Wet etching of ZnO layer (Prasad et al. 2013)
13.	Metal deposition for top split electrodes: RF sputtering: Cr/Au
14.	Patterning **(Lithography M#4)** definition of top electrodes
15.	Wet etching: Cr/Au: formation of top electrodes
16.	**Lithography M#5** for release of beams from front side
17.	DRIE of Si from front side: release of guided beams (developed)
18.	Separating the dies: self-release through DRIE of Si (developed)
19.	Packaging of device in pre-fabricated PCBs (developed)
20.	Wire bonding (on PCB) optimized
21.	Testing and characterization (designed and developed circuitry)

cleaning of the wafers was done in two steps: degreasing and chemical cleaning. After degreasing of the wafers, chemical cleaning was done using piranha. Piranha cleaning of the wafers is done using H_2SO_4 and H_2O_2 in the ratio of 3:1 followed by thorough deionized water (DI) water rinsing. HF and DI water solution (1:100) dip are given to remove the oxide followed by thorough DI rinsing. The 0.6 μm thermal oxide layer is grown at 1000 °C followed by 0.2 μm of LPCVD nitride deposition which acts as mask during TMAH etching.

Photolithography (mask #1) is performed for TMAH opening to perform wet etching. Positive photoresist S1818 is used and CD26 its corresponding developer for masking the desired pattern followed by reactive ion etching (RIE). Wafers are subjected to bulk micromachining of Si in 25% wet TMAH etching at 79 °C giving an etch rate of 20 μm/hr. Dry oxidation of 0.1 μm is performed to act as an insulation between silicon and bottom electrode. Gold sputtering (0.2 μm) followed by patterning (mask #2) and etching is done to form the bottom electrode. A 2-μm-thick ZnO layer using RF magnetron sputtering is deposited followed by patterning (mask #3) and wet etching to form the piezoelectric layer. Gold sputtering (0.2 μm)

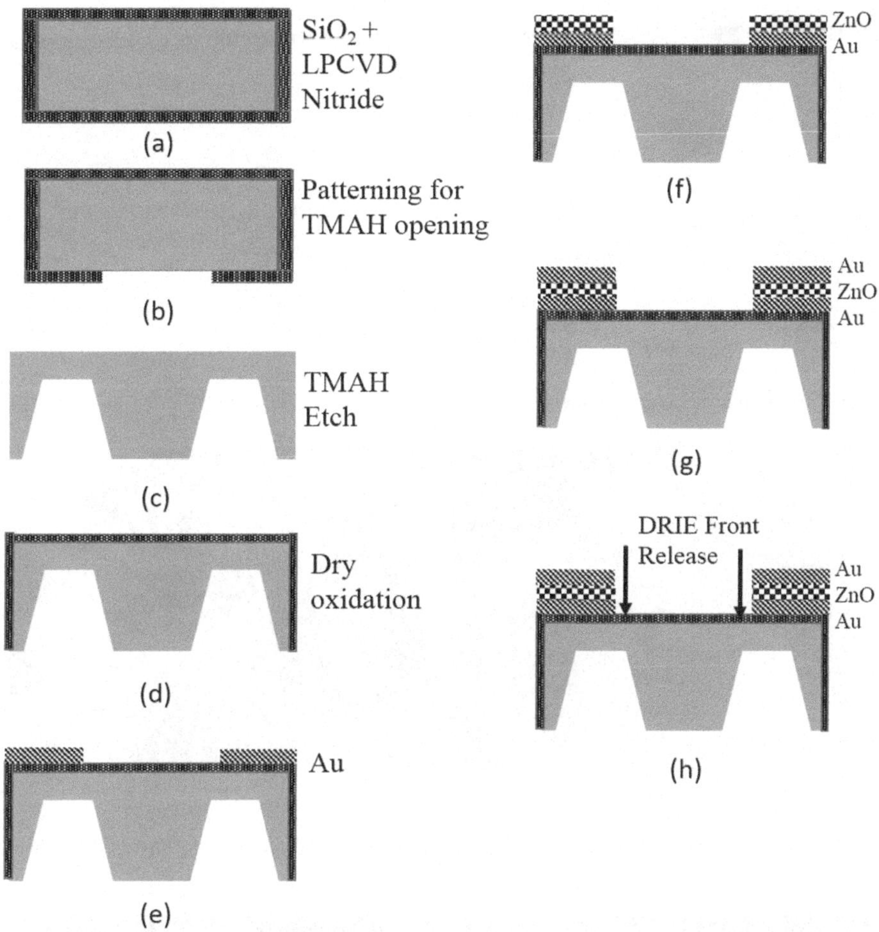

Fig. 5.7 Fabrication flow steps of the device—**a** Thermal oxidation and LPCVD nitride deposition, **b** oxide patterning for pyramidal-shaped seismic mass (mask #1) and etching **c** TMAH etching. **d** Dry oxidation, **e** Cr-Gold deposition and patterning (mask #2), **f** ZnO deposition and patterning (mask #3), **g** gold deposition and patterning (mask #4). **h** Patterning for release of guided beams (mask #5)

followed by patterning (mask #4) and etching is done to form the top electrode. The photolithography (mask #5) is done using positive photoresist S1818, and wafer is patterned from front side. DRIE is performed from the front side of the wafer to release the two-beam and four-beam device.

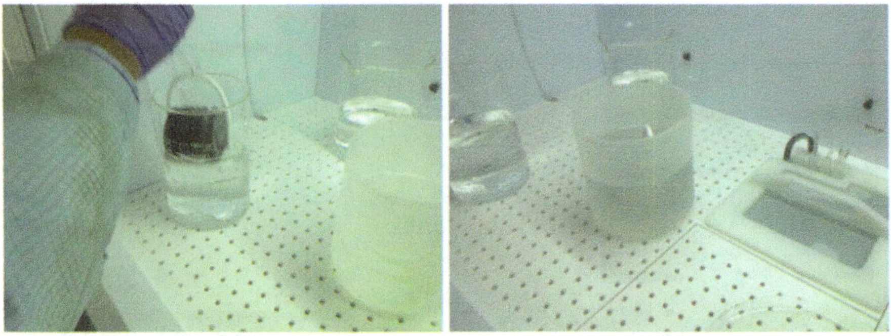

(a) (b)

Fig. 5.8 **a** Piranha cleaning of wafers. **b** HF dip of wafers for oxide removal

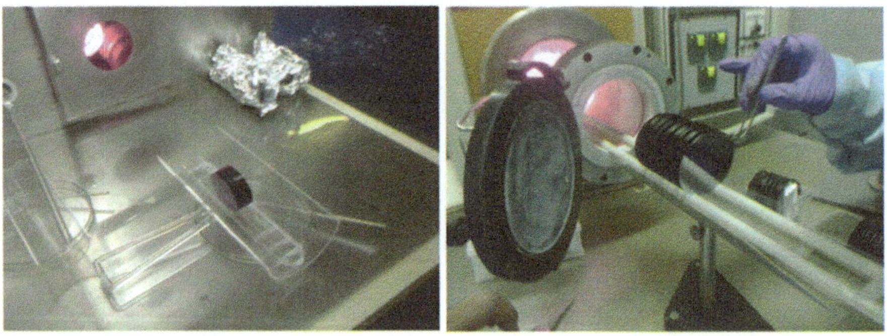

(a) (b)

Fig. 5.9 **a** Wafers loading in thermal oxidation furnance. **b** Wafers loading in LPCVD nitride furnance

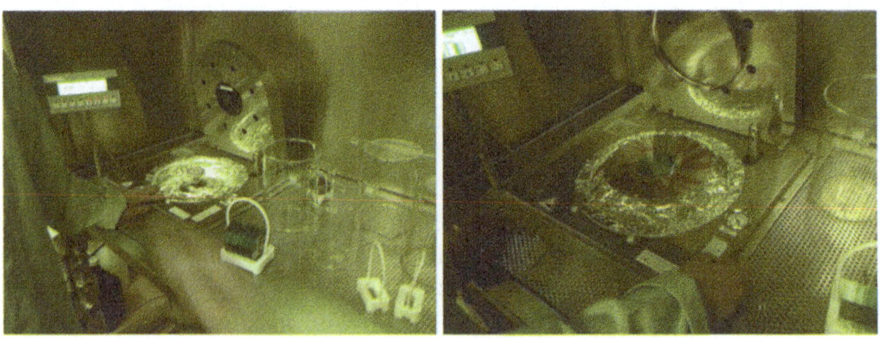

(a) (b)

Fig. 5.10 **a** Wafers after moisture bake for photoresist coating. **b** Positive photoresist S1818 coating process on spin coater

5.4 Detailed Unit Processes

In this section, the unit processes used for the fabrication of the devices are described in detail.

5.4.1 *Cleaning*

Piranha cleaning of the wafers is done on inorganic wet chemical bench using H_2SO_4 and H_2O_2 in the ratio of 3:1 for 15 min followed by multiple times thorough deionized (DI) water rinsing holding wafers vertically in a Teflon boat as shown in Fig. 5.8a. After piranha cleaning, HF and DI water solution (1:100) dip are given in a Teflon beaker to remove the oxide formed due to piranha cleaning followed by thorough DI water rinsing as shown in Fig. 5.8b. The silicon wafer surface becomes hydrophobic after HF dip giving clean silicon surface required for thermal oxidation.

5.4.2 *Thermal Oxidation and LPCVD Nitride*

Thermal oxidation followed by LPCVD nitride deposition is done which act as mask during TMAH etching. The batch of clean wafers is loaded to furnace for thermal oxidation as shown in Fig. 5.9a. Thermal oxidation is done at 1100 °C for 10–60–10 min (dry–wet-dry) giving oxide thickness of ~0.6 μm. Low-pressure chemical vapor deposition (LPCVD) nitride deposition is done immediately after the oxidation process. Wafers are loaded in LPCVD furnace as shown in Fig. 5.9b for nitride deposition. A uniform thickness of nitride is ranging from 0.17 to 0.2 μm which acts as masking layer during TMAH etching.

5.4.3 *Lithography (Mask #1) for Pyramidal-Shaped Seismic Mass*

Lithography using mask #1 is done for wet TMAH etching to realize the pyramidal-shaped seismic mass using bulk micromachining. The wafers are thoroughly cleaned using piranha cleaning followed by DI water rinsing. The wafers are kept in oven for moisture bake for 30 min at 120 °C. Figure 5.10a shows the lithography spin coating bench where photoresist (PR) is coated. Positive photoresist S1818 is coated on the wafer using recipe from spin coater which gives required PR thickness of 3 μm as shown in Fig. 5.10b. The batch of six wafers is PR coated with the same recipe. After PR coating, the wafers are soft baked (pre-baked) for 30 min at 90 °C. The wafer is aligned with mask #1 with the SUSS MicroTec lithography machine as shown in Fig. 5.11a.

After proper alignment, the ultraviolet (UV) exposure is done using the SUSS Micro Tec. UV exposure of 6 s is done for three times as seen from Fig. 5.11b.

After the exposure, the pattern is developed using developer CD26 (shown in Fig. 5.12a) corresponding to the positive photoresist S1818. The developing process is done by dipping the wafers in the developer for 50 s followed by rinsing in DI water as shown in Fig. 5.12b. This process is repeated for the batch of six wafers. After thorough rinsing, the wafers are nitrogen dried and are hard baked (post baked) for 30 min at 120 °C.

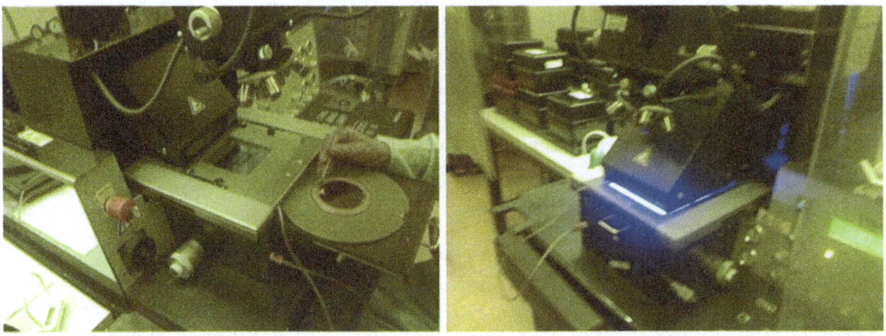

(a) (b)

Fig. 5.11 **a** PR coated wafer alignment with mask #1. **b** Exposure of the wafer through mask #1

(a) (b)

Fig. 5.12 **a** Developer CD26 (left) and two DI water beakers. **b** Developing UV exposed PR coated wafer in CD26 developer

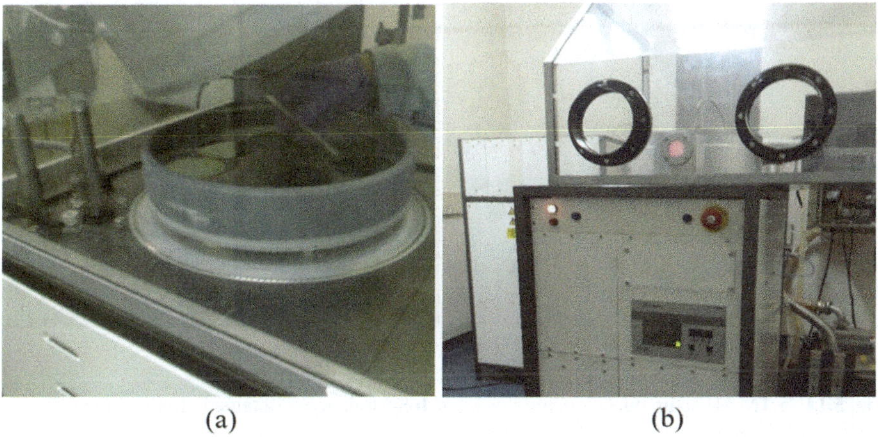

Fig. 5.13 **a** Wafers loading in RIE machine and **b** plasma etching during the process

5.4.4 Reactive Ion Etching (RIE) for Oxide and Nitride Etch

After the lithography with mask #1 is done, the oxide (0.6 μm) and nitride (0.2 μm) are selectively etched using reactive ion etching. The white area as shown in mask #1 (Fig. 5.3a) is TMAH opening for wet etching from where oxide and nitride are etched using RIE. Process parameters are gas flow SF6 (2.1 sccm), CF4 (34.9 sccm), power 100 W. Figure 5.13a shows the loading of the six wafers into RIE. Figure 5.13b shows the plasma generation during etching process. The plasma etches the unprotected PR area.

5.4.5 Bulk Micromachining of Silicon for Pyramidal-Shaped Seismic Mass

After RIE of nitride and oxide, PR removal for the wafers is done by rinsing in acetone first and again boiling in acetone for 10 min (boiling temperature of acetone is 56 °C). After DI water rinse and nitrogen drying the wafers are put in TMAH setup as shown in Fig. 5.14a, at a temperature of 90 °C first for 4 h and using profilometer the thickness measured is 140 μm giving an etch rate of 35 μm per hour. Second interval for TMAH 90 °C is done, and 256.3 μm is etched measured using profilometer. TMAH etch for wafers done in the third interval for 3.5 h etching 292 microns in fourth interval for 30 min and 8 min in fifth interval reaching a final thickness of 345–352 μm. Figure 5.14b shows the pyramidal-shaped seismic mass obtained after TMAH etching. The diaphragm thickness left is 20–25 μm.

After TMAH etching, the wafers are undergone piranha cleaning (mild). Concentrated HF dip is given to the wafers for oxide and nitride removal as shown in

(a) (b)

Fig. 5.14 **a** TMAH setup during etching process. **b** Seismic mass obtained after etching

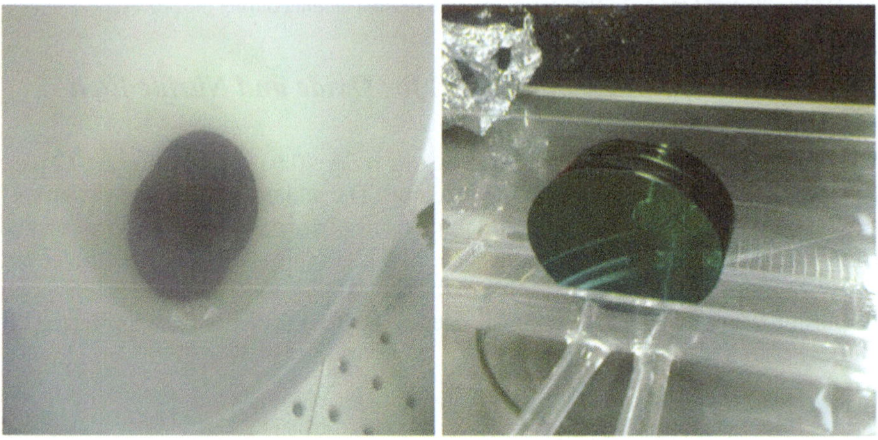

(a) (b)

Fig. 5.15 **a** HF dip of wafers for oxide and nitride removal and **b** dry oxidized wafers

Fig. 5.15a. After cleaning, dry thermal oxidation of ~0.1 micron is grown on the wafers shown in Fig. 5.15b which act as an insulation layer between silicon and electrodes on the front side of the wafer.

5.4.6 Gold Sputtering and Bottom Electrode Patterning (Mask #2)

After dry oxidation, piranha cleaning of the wafers is done followed by DI water rinsing and N_2 drying. Sputtering of Chrome/gold (0.02/0.2 μm) is done on the front side of the wafers. Mask #2 is used to pattern the wafers to form the bottom electrodes. Same lithography procedure is followed as described above. The PR of thickness 1.3 μm is coated. The TMAH etching is done on the bottom side of the wafer, and the gold deposited is on the front side of the wafer; this requires back side alignment of the wafers from mask #2. This is done by aligning the front side from the rectangle-shaped alignment marks specifically designed to sustain TMAH etching as shown in Fig. 5.16. The UV exposure is given for 6 s two times and is developed as shown in Fig. 5.16b.

The wafers are hard baked, and then wet etching of the patterned wafer is done. The gold etchant is used to etch the gold, and the PR is removed and then chromium is etched in chrome etchant. The wafers are rinsed in DI water after etching and put in acetone for PR removal as shown in Fig. 5.17a. Figure 5.17b shows the patterned bottom electrode wafer.

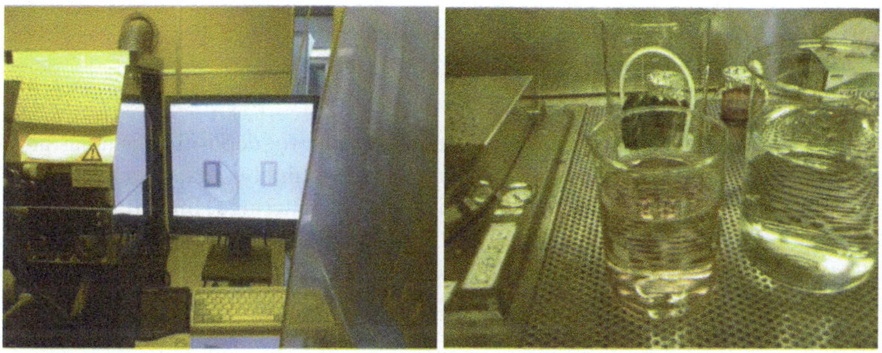

(a) (b)

Fig. 5.16 **a** Back side alignment for bottom electrode patterning. **b** Developing pattered wafers

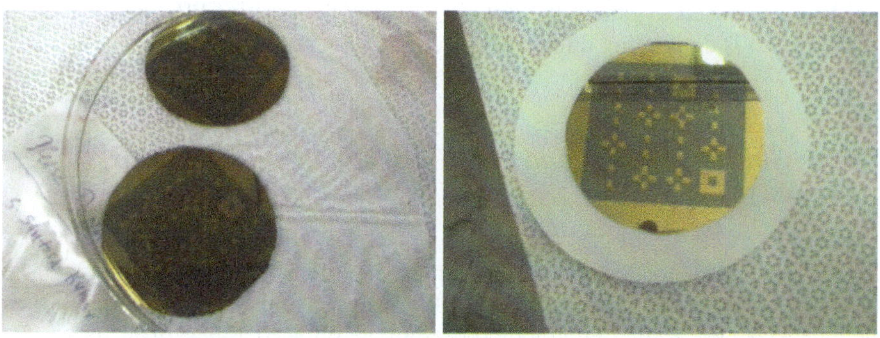

(a) (b)

Fig. 5.17 **a** PR removal after chrome/gold etching of wafers. **b** Patterned bottom electrode wafer

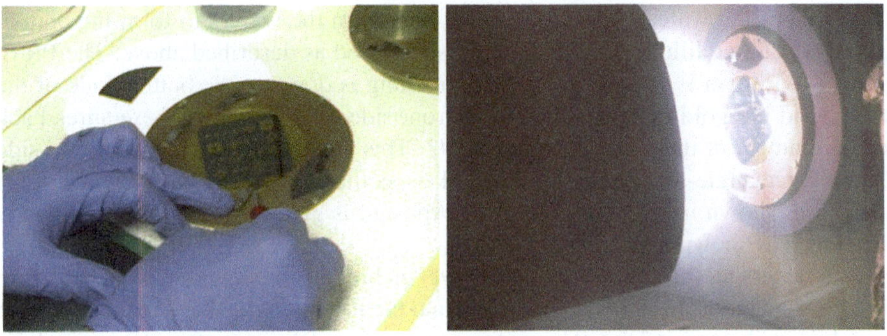

(a) (b)

Fig. 5.18 **a** Wafer mounting on holder of ZnO sputtering machine. **b** ZnO deposition in process

5.4.7 Piezoelectric Layer (Zinc Oxide) Sputtering and Patterning (Mask #3)

Zinc oxide (ZnO) is the piezoelectric material selected for deposition as it has good piezoelectric properties and is CMOS compatible. A highly c-axis-oriented piezoelectric film is deposited using reactive magnetron sputtering (Hind High Vacuum, Pvt. Ltd.). The following parameters are used for ZnO deposition: power: 450 W, gases used are argon: 26 sccm and oxygen: 39 sccm, pressure: 26 mtorr giving a deposition rate of 0.5 μm/hr. The total deposition time is 4 h giving a thickness of 2 μm. Wafer having bottom electrodes is cleaned by degreasing process. Wafers are boiled in trichloroethylene (TCE) for 15 min, followed by acetone boiling for 15 min and then isopropylalcohol (IPA). After thorough rinsing by DI water and nitrogen drying, the wafer is mounted on holder of the sputtering machine as shown in Fig. 5.18a. Figure 5.18b shows the ZnO deposition in process.

Mask #3 is used to pattern the ZnO layer using the positive PR, and the etching of ZnO layer is done using dilute HCl solution.

5.4.8 Gold Sputtering and Top Electrode Patterning (Mask #4)

Degreasing is done to clean the wafers. Sputtering of chrome/gold (0.02/0.2 μm) is done on the front side of the wafers. Mask #4 is used to pattern the wafers to form the top electrodes. Same lithography procedure is followed as described above. The PR of thickness 1.3 μm is coated. The gold etchant is used to etch the gold, and the PR is removed and then chromium is etched in chrome etchant. Figure 5.19a shows the patterned top electrode wafer. The zeta image of top electrode pattern is shown in Fig. 5.19b.

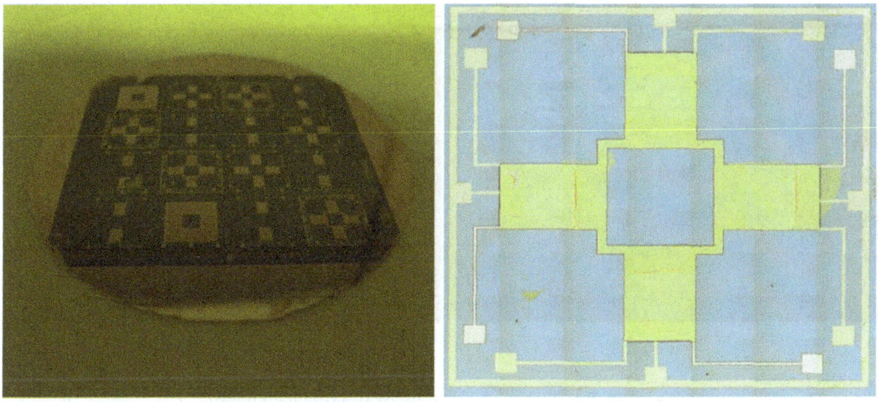

(a) (b)

Fig. 5.19 **a** Patterned top electrode wafer and **b** zeta image of the device from the front showing the top and bottom electrode patterns

5.4.9 Lithography Mask #5 and Beam Release Using DRIE

The two-beam and the four-beam devices are released from the front side using deep reactive ion etching (DRIE). Mask #5 is used to pattern the front side of the wafer for DRIE; after lithography, the wafer is stick to six-inch wafer as the DRIE machine takes six-inch samples. Figure 5.20a, b shows top and lateral views of the process wafer stick to six-inch wafer.

DRIE is done from the front side of the wafer, after etching around 25 μm of silicon from the front the guided two and four beams are released. Figure 5.21a, b shows the top view and lateral view of beam release after DRIE.

After DRIE, positive photoresist is removed in acetone shown in Fig. 5.22a, and

(a) (b)

Fig. 5.20 **a** Wafer stick to six-inch wafer after lithography. **b** Lateral view of the wafer

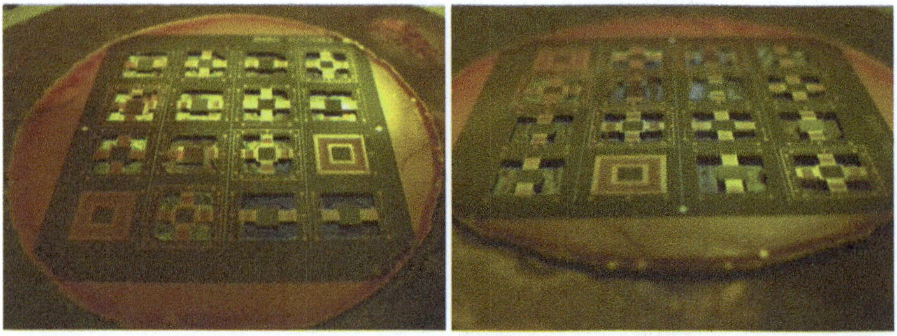

(a) (b)

Fig. 5.21 **a** Beam release after DRIE. **b** Lateral view of the released beams

(a) (b)

Fig. 5.22 **a** PR removal after DRIE. **b** Wafer with released beams

the wafer with released beams is obtained as shown in Fig. 5.22b.

5.4.10 Dicing and Device Release

The devices are diced manually from the grid line designed in mask #1 and mask #5. Figure 5.23a, b shows the snapshot of fabricated guided (a) two-beam and (b) four-beam piezoelectric vibration energy harvester.

(a) (b)

Fig. 5.23 Snapshot of fabricated guided **a** two-beam and **b** four-beam piezoelectric energy harvester

5.4.10.1 Discussions

The fabricated devices are successfully separated and tested for resonance frequency. A laser Doppler vibrometer test is done which is used to determine the resonance frequency of the two-beam and four-beam devices. The device operates in dominant frequency mode and gives displacement in ±Z direction which is discussed in detail in the next chapter. The resonance frequency measured experimentally for two-beam device is 1971 Hz and for four-beam device is 2540 Hz. The cross-sectional SEM image of the beam is shown in Fig. 5.24a. The beam thickness measured is around 25 μm. FEM and analytical calculations at 25 μm thickness give resonance frequency for two-beam device as 1535.2 and 1622 Hz and for four-beam device is 2174.7 and 2294 Hz. The experimental and theoretical results are in close agreement with each other. Due to fabrication imperfections, there is drift in the resonance frequency. The

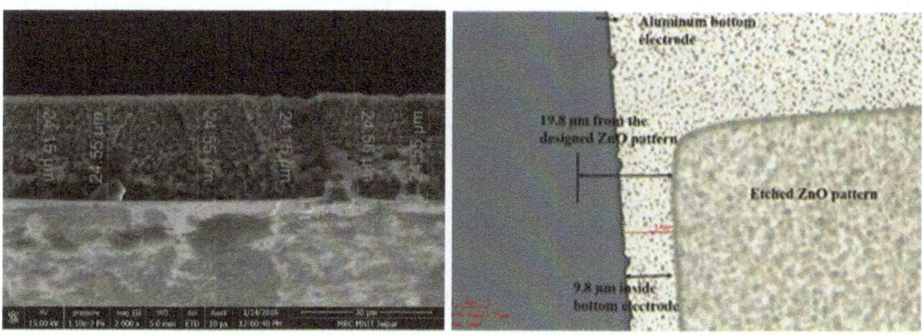

(a) (b)

Fig. 5.24 **a** Cross-sectional SEM image of the beam. **b** Zeta image of the ZnO pattern after etching

resonance frequency especially for the four-beam device is high for energy harvesting applications because ambient vibrations are of low frequency up to 2 kHz. Therefore, the thickness of the beam must be reduced so that the device can have lower resonance frequency.

Figure 5.24b shows the zeta image of the ZnO pattern after etching. It can be seen that 19.8 μm of lateral etching of ZnO occurs while etching 2 μm thickness of ZnO layer. This results in poor connection of bottom and top electrode. The fabrication process is optimized to reduce the beam thickness and for proper passivation between bottom and top electrodes.

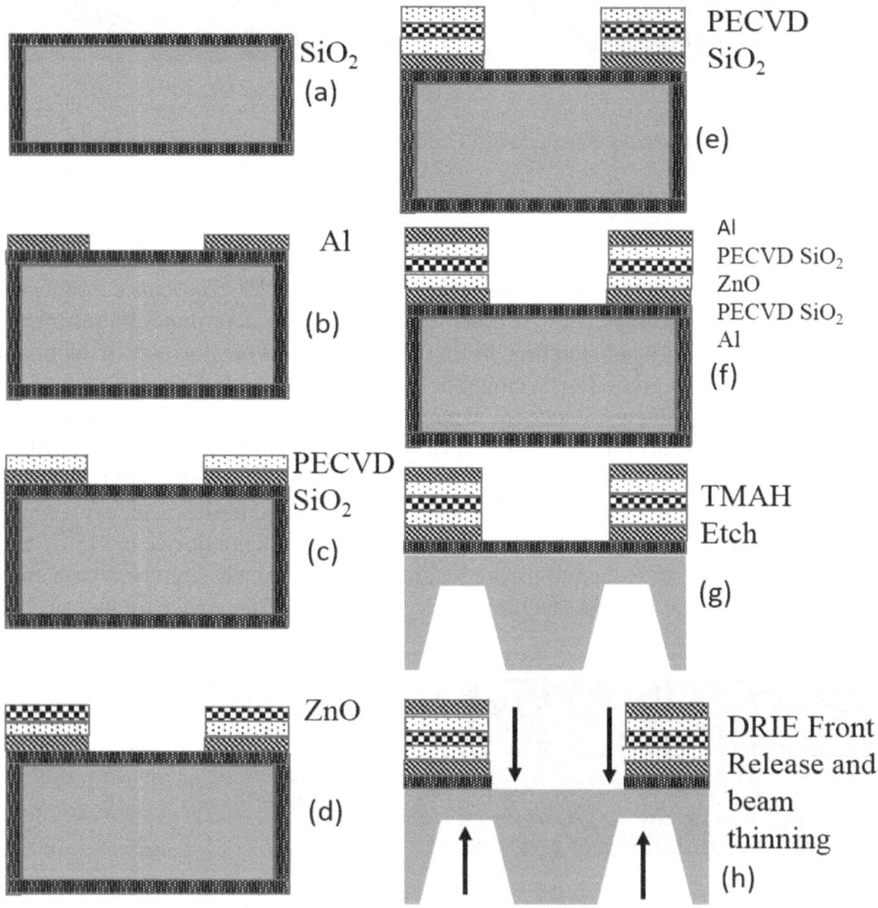

Fig. 5.25 Fabrication steps for device—**a** Thermal oxidation, **b** aluminum deposition and patterning (mask #1), **c** PECVD SiO_2 deposition, **d** ZnO deposition and patterning (mask #2), **e** PECVD SiO_2 deposition, **f** aluminum deposition and patterning (mask #3), **g** oxide patterning for pyramidal-shape seismic mass (mask #4) and TMAH etching, **h** patterning for guided beams (mask #5), DRIE front release and beam thinning from back side and dicing

5.5 Fabrication Process Optimization and Device Fabrication (Batch 2)

The thickness of the beams must be reduced in order to lower the resonance frequency of the device. During device fabrication, the TMAH etching was done after first lithography, due to which the thickness of the beams cannot be reduced after a certain value because the handling of the wafer becomes difficult in the subsequent steps. Therefore, to overcome this issue first the front side patterning of bottom electrode, piezoelectric layer and top electrode is done; then TMAH etching is done followed by DRIE which resulted in a controlling the beam thickness in a more effective way. Secondly, to improve the step coverage plasma-enhanced chemical vapor deposition (PECVD) oxide is deposited after bottom electrode patterning and after ZnO patterning which resulted in proper charge generation and collection and prevented the ZnO layer from environmental deterioration. The optimized fabrication flow and related process steps are discussed.

5.5.1 Fabrication Flow

Process sequence optimization for fabrication of guided two-beam and four-beam piezoelectric energy harvester device is shown in Fig. 5.25. Front side processing is done first, followed by TMAH and DRIE process.

5.5.2 Fabrication of Structure on Front Side of Wafer

At front side, after oxidation, aluminum metal having thickness of 350 nm is deposited using e-beam technique. Lithography using mask #2 is done to form the bottom electrode pattern. Positive photoresist S1818 is used for wafer coating, CD26 is the corresponding developer used to mask the bottom electrode metal pattern followed by wet etching using Al etchant. A thick 0.5 μm SiO_2 layer is deposited using PECVD technique which prevents the diffusion of piezoelectric layer into the metal. Deposition of 2.5 μm thick ZnO layer using RF magnetron sputtering is done which acts as piezoelectric layer responsible for charge generation. The following environmental parameters were kept during deposition, power 400 W, gas composition as 40% Ar + 60%O_2, pressure 30 mtorr and deposition rate 0.5 μm/hr. A maximum thickness of ZnO is selected as 2.5 μm, because after this thickness piezoelectric properties get saturated (Prasad et al. 2013). Second lithography using mask #3 is done to pattern the ZnO film. Dilute Hcl solution in 1% concentration is used to etch the ZnO layer. A thick 0.5 μm of PECVD SiO_2 layer is deposited over patterned ZnO film for passivation and to protect from environmental deterioration.

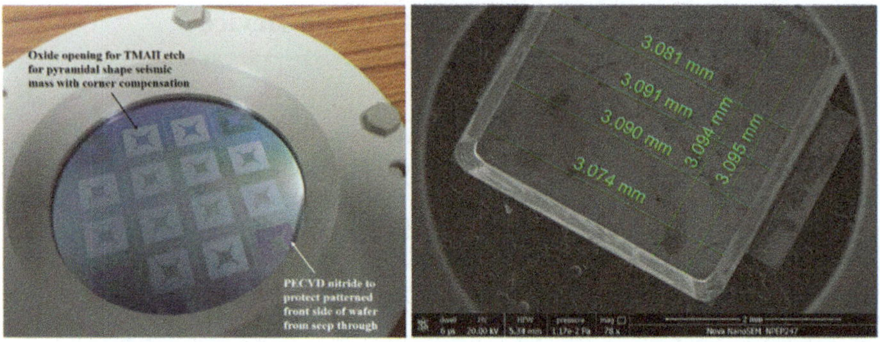

(a) (b)

Fig. 5.26 **a** Snapshot of the front side protection of the processed wafer with Zig for TMAH etching and **b** bottom side SEM image of the seismic mass having smooth surface and perfect edges

Aluminum metal of 1 μm thickness is sputtered over SiO_2 layer. Mask #4 is used to pattern the Al layer to form the split electrodes.

5.5.3 Bulk Micromachining of Silicon for Pyramidal-Shaped Seismic Mass

As discussed, the wet bulk micromachining is done after front side patterning to have a better control over etch depth and also handling of wafers becomes easier resulting in higher yield. To protect the patterned front side of the wafer from the wet etchant, a 3-inch Zig (by AMMT) is used which completely protects the front side of the wafer during wet etching shown in Fig. 5.26. Fourth lithography is done using Mask # 1 to pattern thermal oxide layer followed by reactive ion etching for opening the desired dimension of the thermal oxide layer for TMAH etching of silicon. After this step, the wafers are subjected to bulk micromachining process of Si in 25% wet TMAH etching solution at a temperature of 79 °C which gives an etch rate of 20 μm/hr. A heavy pyramidal-shaped seismic mass having smooth surface and perfect edges due to corner compensation is realized as shown in Fig. 5.26b.

5.5.4 Guided Beam Release and Thinning Using Deep Reactive Ion Etching (DRIE)

In final stage of the device fabrication, the four-beam and two-beam structures are released. Dry etching is done through deep reactive ion etching which is a preferred process during beam releasing of the device. Fifth lithography is done using dark

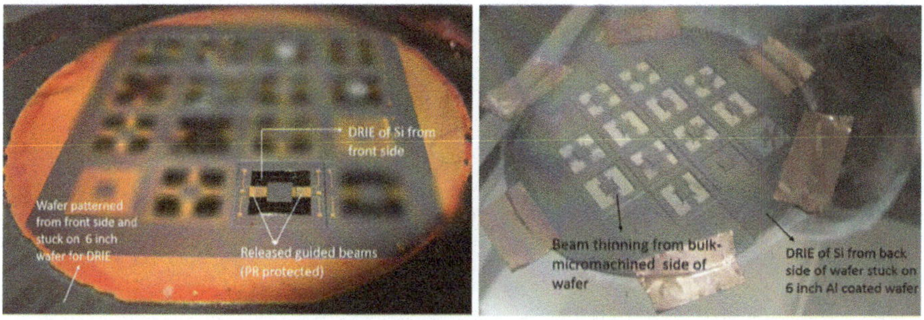

(a) (b)

Fig. 5.27 **a** Snapshot of wafer after DRIE etch of Si from front side. **b** Snapshot of bulk micromachined side of the wafer in process during beam thinning process

field Mask #5 on the front side of the wafer to define the beam structure. The process wafer is stuck on a six-inch silicon wafer while performing DRIE from the front side of the wafer. Figure 5.27a shows the released two-beam structure from the front side. To effectively control the beam thickness, beam thinning is done from the back side to achieve a beam thickness of less than or equal to 10 μm as shown in Fig. 5.27b. The process of beam release from the front side followed by beam thinning from the back side using DRIE was optimized in several different runs as it is the most critical step in defining the beam thickness and therefore the resonance frequency of the energy harvester. The process is done with an Alcatel AMS100 DRIE machine. The process parameters for DRIE are as follows: Pressure = 4.8×10^2 mbar, Source RF1 = 1000 W, Chuck RF2 = 20 W. SF6 = 600 sccm. Front side beam release is done by etching of the silicon with the aforesaid parameters for 8 min, which etches Si up to a depth of ~25 μm (front side) giving an etch rate of 3.125 μm/minute. Back side blanket etching is done of Si wafer using the same recipe but in separate intervals of 2 min each.

5.5.5 Dicing and Device Release

The devices are self-diced by etching the grid lines during the etching process. A small press along the etched grid lines resulted in the releasing of the device. Figure 5.28a, b gives the snapshot of fabricated guided (a) two-beam and (b) four-beam piezoelectric energy harvester.

The device fabrication is completed successfully, and both two-beam and four-beam devices are successfully diced for packaging and testing.

Table 5.3 gives the comparison of the two-beam and four-beam devices before and after fabrication process optimization. It can be seen that the thickness of the beam is reduced from ~25 to ~12 μm which is 52% less. This reduction in the thickness gives a significant reduction in the resonance frequency of the devices. The resonance

(a) (b)

Fig. 5.28 Snapshot of fabricated guided **a** two-beam and **b** four-beam piezoelectric energy harvester

Table 5.3 Comparison for two-beam and four-beam devices

Parameter	Designed (two-beam/four-beam)	Fabricated (two-beam)	Fabricated (four-beam)
Beam length	2500	2490	2470
Beam width	2000	1959	1978
Seismic mass (top)	3500 × 3500	3552 × 3548	3532 × 3517
Seismic mass (bottom)	3033 × 3033	3081 × 3094	3090 × 3095
Seismic mass (height)	330	336	335.8
Beam (pre-thinning)	20	25.63	25.23
Beam (post-thinning)	10	13	12
Frequency (pre-thinning)	1535/2174 Hz	1971.9 Hz	2540.6 Hz
Frequency (post-thinning)	483/519 Hz	466 Hz	515 Hz

frequency after optimization comes to around 500 Hz which is suitable for vibration energy harvesting.

5.6 Design of the PCB and Packaging

After dicing and separating the chips, they have to be tested in different ways. For resonance frequency determination, an LDV is used. For LDV testing, the bare chip is used on a piezoelectric exciter. The devices have to be further tested for their outputs in vibration ambiance. For vibration testing, the device is mounted on a vibration table (shaker) and we have to take out the connections from the output pads of the device. Therefore, a special kind of PCB was designed having gold plated pads suitable for wire bonding. PCB designed for both two-beam and four-beam devices is shown in Fig. 5.29a. Figure 5.29b shows the device mounted on PCB package using nonconducting epoxy and curing it at 130 °C for 30 min. The device is mounted on PCB and wire bonded from the chip to the PCB. After this, Perspex caps are used to seal the device so that it can be tested in ambient environment. To connect the device to the analyzer for vibration bench testing, connectors are to be connected for proper testing of devices. Therefore, a special kind of PCB attachment was designed with a built-in connector and attached to the PCB so that the coaxial cable can be connected to the analyzer. The device packaged in this way has to be mounted on the vibration table, and output of the device can be directly taken from the coaxial output connectors as shown in Fig. 5.30.

Therefore, the devices are ready to be tested for vibration shaker testing, laser Doppler vibrometer testing and also for structural characterization which is presented in the next Chap. 6.

(a) (b)

Fig. 5.29 **a** Snapshot of PCB package. **b** Device mounted on a PCB package

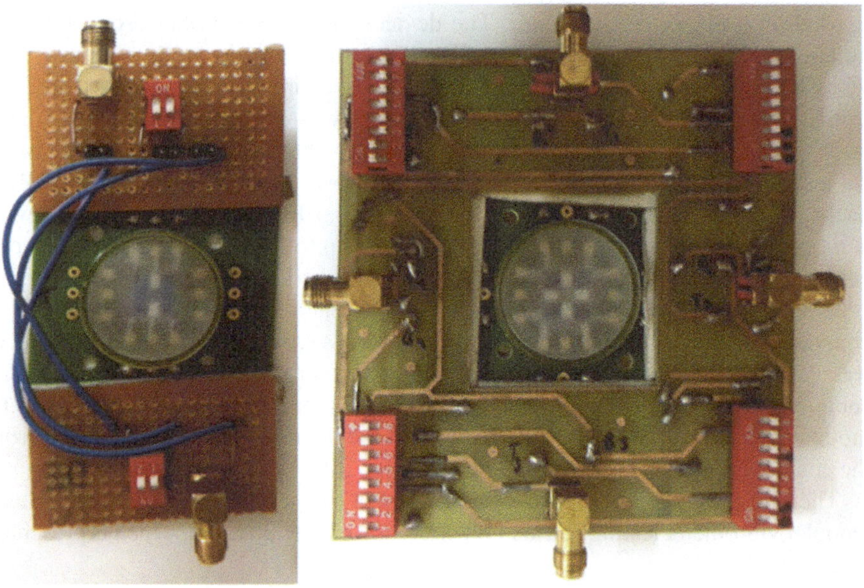

(a) (b)

Fig. 5.30 a, b Devices mounted on PCB with connectors to connect the devices to the vibration analyzer

References

Liu C (2012) Foundations of MEMS. Pearson Education India

Mukhiya R, Bagolini A, Margesin B, Zen M, Kal S (2006) <100> bar corner compensation for CMOS compatible anisotropic TMAH etching. J Micromech Microeng 16(11):2458

Prasad M, Sahula V, Khanna VK (2013) Design and fabrication of Si-diaphragm, ZnO piezoelectric film-based MEMS acoustic sensor using SOI wafers. IEEE Trans Semicond Manuf 26(2):233–241

Chapter 6
Testing and Characterization of Guided Beam Piezoelectric Energy Harvester

Abstract This chapter presents testing and characterization for guided two-beam and four-beam P-VEH. Structural characterization for the device has been done using XRD, SEM and AFM to determine the phase orientation of materials, surface morphology, topological view and surface roughness of the deposited films. A laser Doppler vibrometer has been used to determine the resonance frequency of the devices. Resonance frequency of the two-beam device comes at 466 Hz, and for the four-beam device, resonance frequency is 515 Hz. Vibration shaker test using SPEKTRA SE-10 is performed to measure the output potential generated at a given acceleration on two-beam and four-beam devices. In the low-frequency domain (10–1000 Hz), two-beam and four-beam devices exhibit net sensitivities of 2.2784 $mV/m/s^2$ and 4.09272 $mV/m/s^2$, respectively.

Keyword SEM · XRD · AFM · Vibration Bench · LDV · Sensitivity

In this chapter, an elaborate characterization of both kinds of fabricated devices two-beam and four-beam P-VEH has been done. Some of the characterization has been carried out during the fabrication process itself, that is, the wafer-level characterization, required to ensure the correct dimensions of different structures such as beams and suspended mass. In order to obtain the optimum output from the device, it is necessary to characterize the device at each level including the fabrication steps. It is quite difficult to obtain the correct dimensions of different structures in wet bulk micromachining; still very close dimensions have been achieved.

The elaborate characterization includes structural, material, morphological, topological, dynamic and electrical characterization of the device. Different tools have been utilized to carry out all these characterizations. Structural characterization is done in which, phase orientation of materials is done using X-ray diffraction (XRD), surface morphology is obtained using scanning electron microscope (SEM), and topological view is obtained with atomic force microscopy (AFM). Dynamic characterization is performed using a laser Doppler vibrometer (LDV) to experimentally obtain the resonance frequency of the two-beam and four-beam devices. Finally, vibration bench testing is done for measurement of electric potential and sensitivities of the two-beam and four-beam devices.

S. Saxena et al., *Design and Development of MEMS based Guided Beam Type Piezoelectric Energy Harvester*, Energy Systems in Electrical Engineering,
https://doi.org/10.1007/978-981-16-0606-9_6

6.1 Structural Characterization

Structural characterization for the device has been performed at Materials Research Center at MNIT, Jaipur, India. XRD of the device is done using Powder XPERT PRO from PANalytical to determine the phase orientation of the deposited zinc oxide (ZnO) and gold (Au) materials. The XRD pattern of the device is shown in Fig. 6.1a. There are clearly two dominant peaks seen from Fig. 6.1a; The first peak

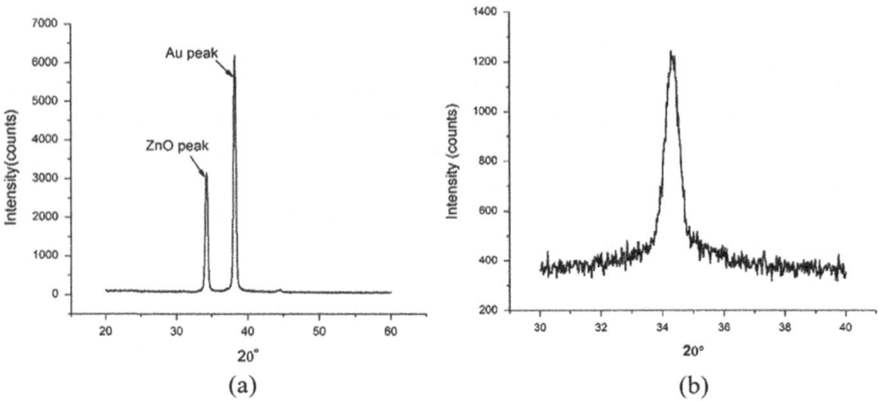

Fig. 6.1 **a** XRD pattern of the device **b** XRD pattern of ZnO deposited on a bare silicon wafer

comes at 34.4° which represents the peak of ZnO material deposited and the second peak comes at 38.18° which represents peak of gold (Au) material deposited to form the bottom and top electrodes.

Figure 6.1b gives XRD pattern of deposited ZnO thin film on a bare silicon wafer. Thickness of ZnO is kept 2.5 μm because above this thickness piezoelectric properties get saturated (Prasad et al. 2013). It is observed that the growth of the film is highly oriented along the c-axis normal to the substrate because there only one peak corresponding to (002) reflection of Wurtzite phase of ZnO. At an angle of $2\theta = 34.4°$, the ZnO film exhibits a strong peak, which indicates that the ZnO film grains have good orientation exhibiting good piezoelectric properties.

NOVA NanoFESEM 450 from FEI is used to take SEM images of devices. Figure 6.2 shows the SEM image of the bottom view of pyramidal shape seismic mass. The dimensions of the seismic mass from the bottom are found to be 3084 × 3094 μm as seen from SEM image which is similar to designed dimensions of 3083 × 3083 μm. The corner edges of the seismic mass are smooth due to corner compensation structures designed for TMAH etching.

Figure 6.3 shows the SEM image of the top view of pyramidal shape seismic mass. The dimensions of the seismic mass from the top are 3554 × 3353 μm as seen from SEM image which is similar to designed dimensions of 3500 × 3500 μm. The designed dimensions of the proof mass are obtained in device fabrication.

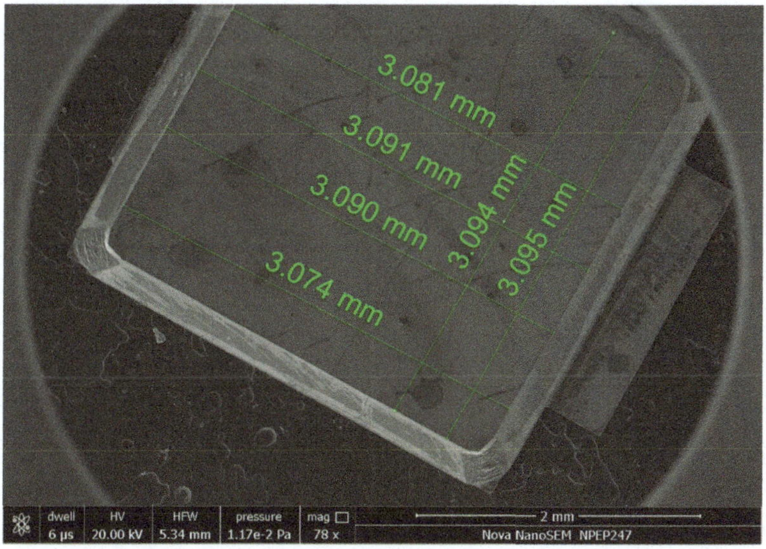

Fig. 6.2 SEM image of the bottom view of the pyramidal shape seismic mass

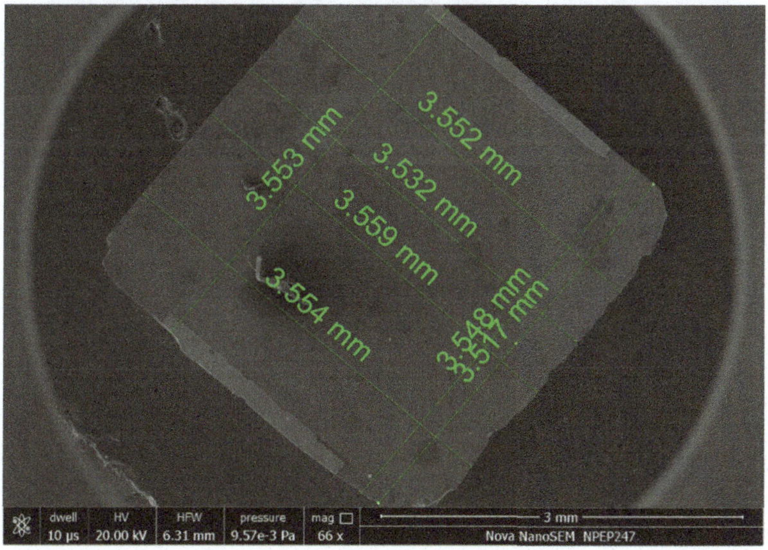

Fig. 6.3 SEM image of the top view of the pyramidal shape seismic mass

Figure 6.4 shows the SEM image of the beam with bottom electrode. The dimensions of the fabricated bottom electrodes are 2543 × 1978 μm which are quite close to the designed dimensions of 2500 × 1980 μm.

Fig. 6.4 SEM image of the beam with bottom electrode

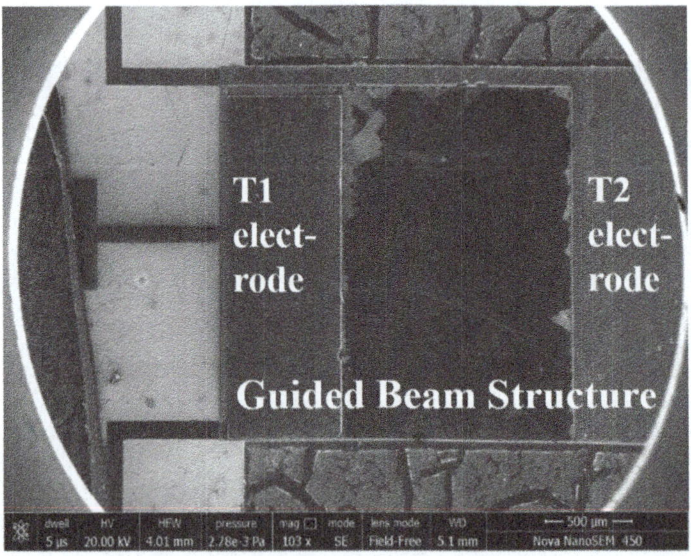

Fig. 6.5 SEM image of the split electrodes each having dimensions of 675 μm

Figure 6.5 shows the SEM image of the beam with bottom and top split electrodes. The dimensions of the fabricated split electrodes are length 675 μm and width 2000 μm.

Figure 6.6 shows the SEM image of the interface of the split electrode and the piezoelectric layer.

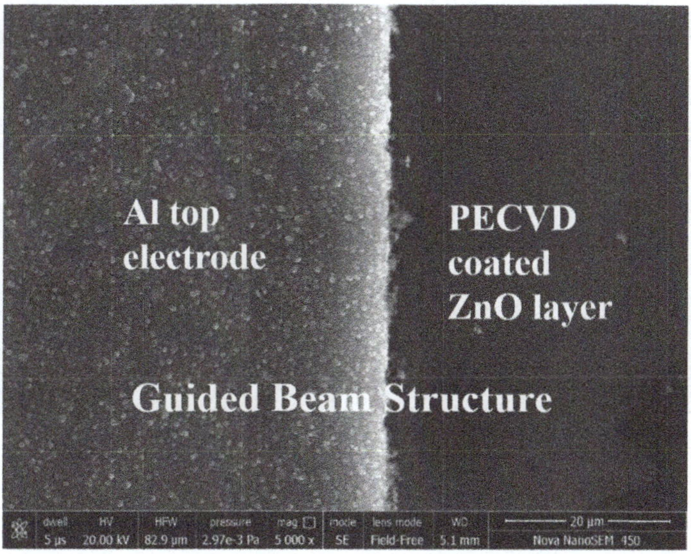

Fig. 6.6 SEM image of the interface of the top electrode and the piezoelectric layer

The AFM imaging was carried by the Multimode 8.0 instrument (Bruker, Germany) in the ScanAsyst mode on the different frame size as 2, 5 and 10 μm. The images infer that a smooth coating with homogeneous texture has been achieved in deposition for gold electrodes as well as ZnO.

Figure 6.7a, b gives the AFM images of the gold metal deposited for bottom and top electrodes.

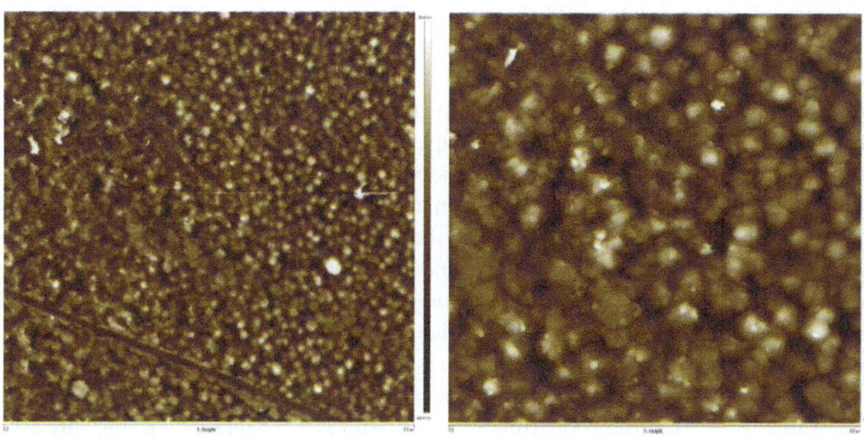

(a) (b)

Fig. 6.7 **a** and **b** AFM images of the gold metal deposited for electrodes

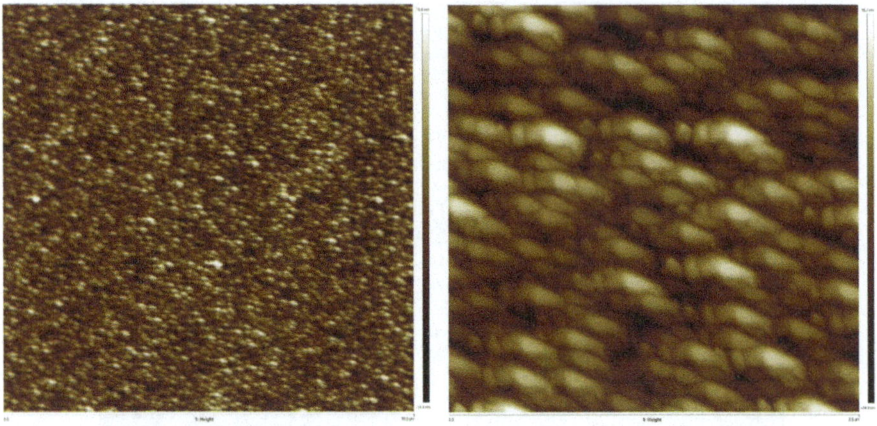

(a) (b)

Fig. 6.8 **a** and **b** AFM images of deposited 2-μm-thick ZnO layer

Figure 6.8a, b shows the AFM of the deposited 2.5-μm-thick ZnO layer. The ZnO layer deposited is highly c-axis oriented as seen from AFM images.

6.2 Dynamic Characterization Using Laser Doppler Vibrometer (Batch1 Devices)

Dynamic performance of the guided two-beam and four-beam piezoelectric energy harvester is performed using a laser Doppler vibrometer (LDV). Resonance frequency of the energy harvesting device is the key parameter as the device should resonate with the ambient vibration frequency for maximum power generation. Laser Doppler vibrometer is used to experimentally calculate the resonance frequency, bandwidth and the quality factor of the device. PolytecMSA-500 MicroSystem Analyzer (LDV), an optical transducer is used for determining the vibration velocity and displacement at a measurement position of the device (Sharma et al. 2015). The dynamic parameters are calculated by sensing the frequency shift of backscattered light from a moving surface focused on the device. The object scatters or reflects light from the laser beam, and the Doppler frequency shift is utilized to measure the component of velocity, which lies along the axis of the laser beam. The signal generator of the MSA-500 generates a broadband excitation signal called as periodic chirp, an AC signal to excite all frequencies in a range simultaneously, and then the fast Fourier transform (FFT) technique is used to measure all of these frequencies at the same time. Maximum instrument bandwidth is 1.5 MHz which measures the velocity of vibration structure. The LDV experimental setup (at CSIR-CEERI) is used to perform dynamic characterization for the guided two-beam and four-beam piezoelectric energy harvester is shown in Fig. 6.9. A piezoelectric disk is used to excite the device; a signal is applied to the disk for base excitation to excite all modes in the band for the device as shown in Fig. 6.10. The scan is performed around the center of the proof mass of the device to obtain data points for maximum displacement.

6.2.1 LDV Measurement for Guided Two-Beam Piezoelectric Energy Harvester

Frequency sweep from 0 to 6 kHz is applied to obtain the displacement versus frequency response for the guided two-beam device. Figure 6.11 gives the scan which clearly shows the mechanical resonance of the device in the Z-direction. Figure 6.12 gives the device resonance frequency measured using LDV which comes at 1971.9 Hz and having a bandwidth of 40.6 Hz. The maximum displacement measured at resonance frequency is 49.67 nm. Frequency response curve for the air-damped device clearly shows device operation in the dominant frequency mode. Since the device is mechanically excited using a piezoelectric disk, the electrical effects on spurious modes are eliminated. It can be seen from Fig. 6.12 that secondary mode for the device comes around 4 kHz is suppressed and predominated by first mode. Therefore, the guided two-beam device provides a stable and reliable operation by operating in dominant frequency mode and providing maximum displacement in Z-direction which leads to maximum power generation.

Wideband frequency operation is another important parameter so that the energy harvester can resonate with a range of vibration frequencies for maximum power generation. This guided two-beam piezoelectric energy harvester gives a 3 dB bandwidth of 40.6 Hz. Maximum displacement measured at resonance frequency is 49.67 nm which will generate stress on the two-beams resulting in potential generation.

Guided two-beam device gives a resonance frequency of 1622 Hz when calculated analytically using Eq. (4.1), whereas resonance frequency obtained using FEM comes at 1535.2 Hz. Resonance frequency experimentally calculated using LDV comes at 1971.9 Hz; therefore, the analytical calculation, simulation and experimental results are in close agreement to each other.

6.2.2 LDV Measurement for Guided Four-Beam Piezoelectric Energy Harvester

Frequency sweep from 0 to 8 kHz is applied to obtain the displacement versus frequency response for the device. The frequency response for the energy harvester device is shown in Figs. 6.13 and 6.14, and the device resonance frequency measured with LDV is at 2540.6 Hz and with a bandwidth of 101.9 Hz. The maximum displacement measured at resonance frequency is 32.17 nm.

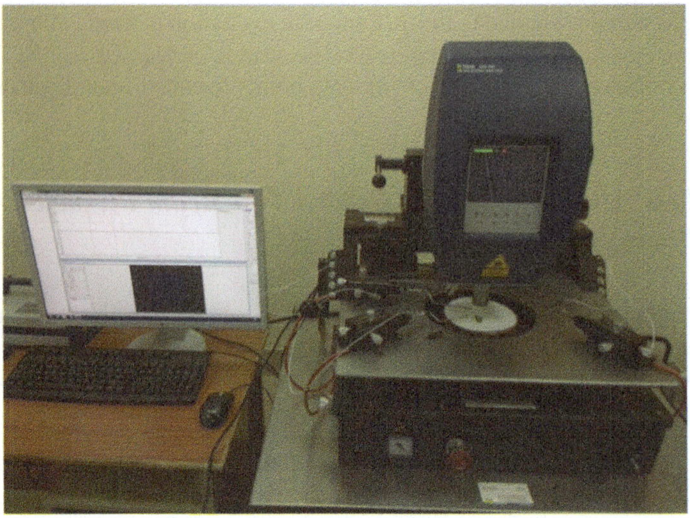

Fig. 6.9 Experimental setup of LDV used to measure dynamic response of the device at CSIR-CEERI

Frequency response curve for the air-damped device clearly shows device operation in the dominant frequency mode. It can be seen from Fig. 6.14 that second mode for the device comes around 5 kHz is suppressed and predominated by the first mode. Therefore, the guided four-beam device provides a stable and reliable operation by operating in dominant frequency mode and providing maximum displacement in Z-direction which leads to maximum power generation. The guided four-beam device gives a 3 dB bandwidth of 101.9 Hz. Maximum displacement measured at resonance frequency is 32.17 nm which will generate stress on the four beams resulting in a potential generation.

Guided four-beam device gives a resonance frequency of 2294 Hz when calculated analytically using Eq. (4.1), whereas resonance frequency obtained using FEM comes at 2174.7 Hz. Resonance frequency experimentally calculated using LDV comes at 2540.6 Hz; therefore, the analytical calculation, simulation and experimental results are in close agreement to each other. There is a slight deviation of the value of the resonance frequency calculated using FEM and using LDV because fabrication precision could not be achieved due to dependency of wet etching on several factors which results in non-uniform thickness of the beam.

6.3 Dynamic Characterization using LDV (Batch 2 Devices)

As discussed earlier, the resonance frequency of the two-beam and four-beam devices is relatively higher in comparison with the freely available frequencies in ambient vibrations in the environment. The fabrication process was optimized as discussed in Sect. 5.5, and devices are fabricated again (batch 2). The fabricated devices are tested using LDV, and the results are discussed.

6.3.1 LDV Measurement for Guided Two-Beam Device

Frequency sweep from 0 to 2 kHz is applied to obtain the displacement versus frequency response for the device. The frequency response for the two-beam device is shown in Fig. 6.15. The device resonance frequency measured using LDV comes at 466 Hz. Frequency response curve for the air-damped device clearly shows device operation in the dominant frequency mode giving displacement in Z-direction. The beam thinning process optimized using DRIE enabled to achieve the lowest resonance frequency so far reported for the guided two-beam device as shown in Table 6.1.

Table 6.1 Comparison of resonance frequency

S. no	Marzencki et al. (2009)	Wang et al. (2012)	Fabricated Device
Resonance frequency (Hz)	1800	694	466

In this case, the device is going to be quite useful to harvest the ambient energy vibrations from mechanical systems such as motors, rotors, engines and turbines of lower frequency range from few hertz up to 500 Hz.

6.3.2 LDV Measurement for Guided Four-Beam Device

Frequency sweep from 0 to 2 kHz is applied to obtain the displacement versus frequency response for the device. The frequency response for the four-beam device is shown in Fig. 6.16. The device resonance frequency measured using LDV comes at 515 Hz. Frequency response curve for the air-damped device clearly shows device operation in the dominant frequency mode giving displacement in Z-direction. The beam thinning process enabled the resonance frequency of four-beam device to reduce from 2540 to 515 Hz which is very suitable for low-frequency ambient vibrations. The overall low-frequency response of the device is improved. Beam thinning process enabled to closely match the resonance frequency of the two-beam and the four-beam device. The resonance frequency of the two-beam device is 466 Hz, whereas for four-beam device is 515 Hz, which will almost give similar low-frequency response. Therefore, the four-beam device and two-beam device have almost same frequency response.

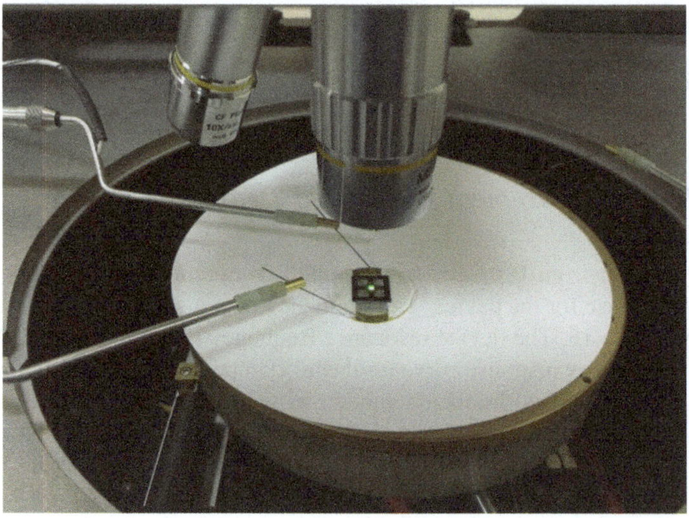

Fig. 6.10 Extended view of the device mounted on the piezodisk for LDV measurement

Fig. 6.11 Snapshot of Scanning LDV measurement result for guided two-beam device. The device operates in dominant frequency mode (±Z-direction)

6.4 Static Capacitance Measurement and Device Packaging

Static capacitance of the device is measured before vibration testing to ensure proper device operation. The static capacitance of the energy harvester device is measured using low-level CV measurement (Keithley, SUSS Microteck) shown in Fig. 6.17. The static capacitance between bottom electrode and electrode T1 (fixed end) is measured as153 pF, whereas static capacitance measured between the bottom electrode and electrode T2 (guided end) is measured as 135 pF. Figure 6.18a shows the

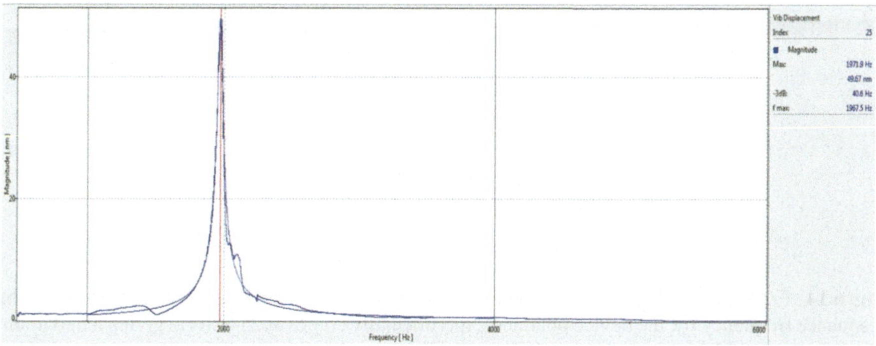

Fig. 6.12 Frequency versus displacement curve for the two-beam device obtained using LDV. The resonance frequency for the device obtained experimentally comes at 1971.9 Hz giving a maximum displacement of 49.67 nm

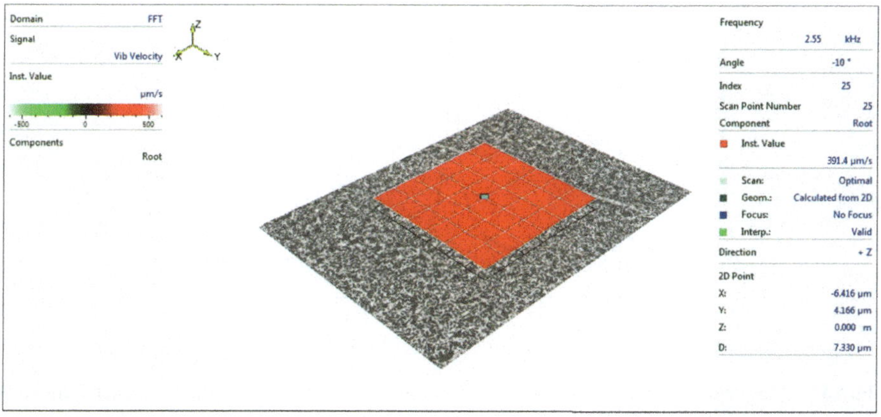

Fig. 6.13 Snapshot of Scanning LDV measurement result for guided four-beam device. The device operates in dominant frequency mode (±Z-direction)

device mounted on a PCB package with non-conducting epoxy, and Fig. 6.18b shows the packaged and sealed device for vibration shaker testing.

6.5 Vibration Shaker Test

Vibration shaker test is performed to measure the output potential generated at a given acceleration on two-beam and four-beam devices. The output potential generated from the individual electrodes T1 (fixed end) and T2 (guided end) as shown in Fig. 6.19 is experimentally measured using a vibration exciter.

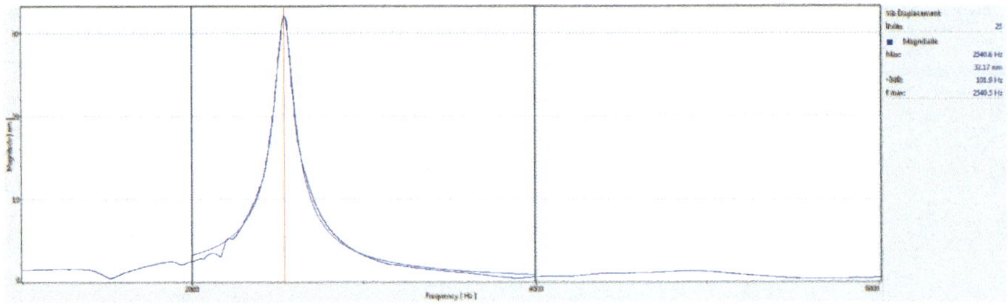

Fig. 6.14 Frequency versus displacement curve for the four-beam device obtained using LDV. The resonance frequency for the device obtained experimentally comes at 2540.6 Hz giving a maximum displacement of 32.17 nm

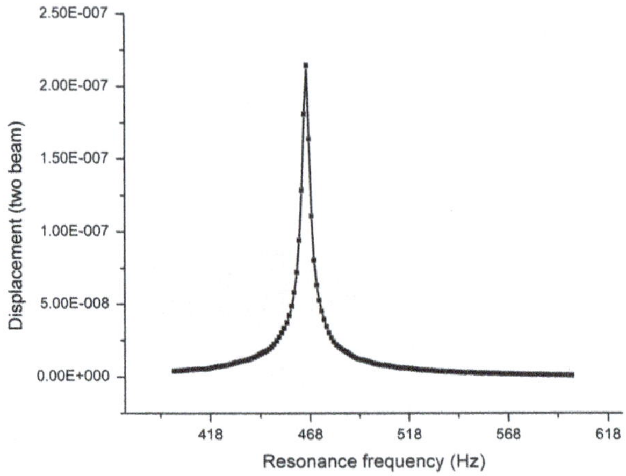

Fig. 6.15 Frequency versus displacement curve for the two-beam device. The resonance frequency for the device obtained experimentally is 466 Hz

The SPEKTRA SE-10 vibration exciter setup is shown in Fig. 6.20. The device is mounted on the shaker, and the data is fed to the analyzer.

The SE-10 vibration shaker has an inbuilt calibrated accelerometer having known sensitivity S1. At a given excitation the accelerometer will generate a voltage V1. The device under test (DUT) mounted as shown in Fig. 6.21 will also generate voltage V2. From Equation 6.1, the S2 of the DUT is obtained. The main advantage of this measurement is that the device sensitivity becomes independent of the applied input vibration.

$$\frac{\mathrm{S1}}{\mathrm{S2}} = \frac{\mathrm{V1}}{\mathrm{V2}} \tag{6.1}$$

The sensitivity of the device S2 is measured for a frequency range from 160 to 1000 Hz as shown in Fig. 6.22. At fixed end from electrode T1, the device sensitivity

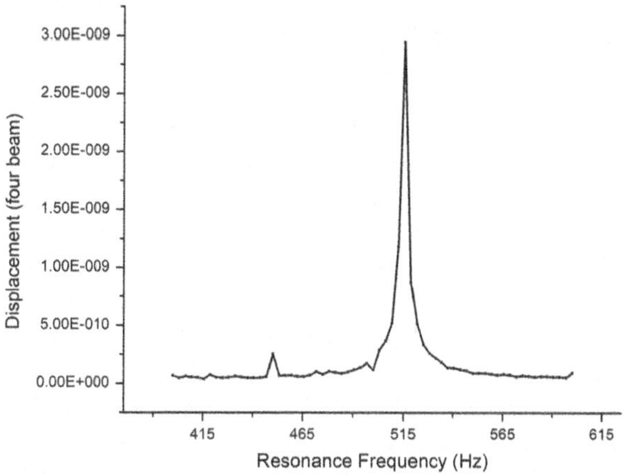

Fig. 6.16 Frequency versus displacement curve for the four-beam device. The resonance frequency for the device obtained experimentally is 515 Hz

measured at 160 Hz is 0.62762 mV/m/s^2 which increases to 0.67484 mV/m/s^2 at 500 Hz. At the guided end from electrode T2, the device sensitivity measured at 160 Hz is 0.83231 mV/m/s^2 and at 500 Hz is 1.5089 mV/m/s^2 which can be seen from Fig. 6.22. It can be seen that both the electrodes give higher sensitivity near the resonance of the device which is 466 Hz. At 800 Hz, the sensitivity of the device measured from T1 electrode is 0.47354 mV/m/s^2 and from T2 electrode is 0.73915 mV/m/s^2. Similarly, at 1000 Hz, the sensitivity of the device measured from T1 electrode is 0.23781 mV/m/s^2 and from T2 electrode is 0.58963 mV/m/s^2. It can be observed from Fig. 6.22 that after resonance frequency (466 Hz) the sensitivity of the device at both the electrodes reduces. Also, the sensitivity of device measured from electrode T2 at guided end is higher than the sensitivity of the device at electrode T1 at the fixed end because the total displacement at the guided end is higher than the fixed end.

Guided four-beam device is tested using SPEKTRA SE-10 vibration exciter setup. Figure 6.23 shows the mounted guided four-beam device on the SE-10 vibration exciter. The sensitivity S2 of the four-beam device is measured for a frequency range from 50 to 1000 Hz and also compared with two-beam device. Figure 6.24 gives the comparison of sensitivity of guided two-beam and four-beam device for the frequency range of (a) 50–1000 Hz (b) 10–50 Hz. At 300 Hz, the two-beam device gives a sensitivity of 2.4066 mV/m/s^2, whereas the four-beam device gives sensitivity of 1.3925 mV/m/s^2. At 500 Hz, near the resonance frequency of both devices the sensitivity for two-beam device is 1.1392 mV/m/s^2 and for four-beam device is 1.0231 mV/m/s^2 as seen from Fig. 6.24a. At 800 Hz, the sensitivity of two-beam device and four-beam device is 0.73915 mV/m/s^{2}and 0.67223 mV/m/s2, respectively. At 1000 Hz, the sensitivity of two-beam device and four-beam device is 0.58963 mV/m/s^2 and 0.56588 mV/m/s^2, respectively.

From Table 6.2, near resonance frequency at 500 Hz two-beam and four-beam devices has sensitivity 1.1392 $mV/m/s^2$ and 1.0231 $mV/m/s^2$. The output is measured for one beam in both two-beam and four-beam devices.

Therefore, the total potential generated for two-beam device is 2.2784 $mV/m/s^2$ (2×1.1392), whereas the total potential generated by four-beam device is 4.09272 $mV/m/s^2$ (4×1.0231) which is 1.814 mV higher than the two-beam device.

Both the devices are tested for very low-frequency range from 10 to 50 Hz. From Table 6.3, it can be seen that two-beam and four-beam give a sensitivity of 1.9777 $mV/m/s^2$ and 2.2968 $mV/m/s^2$, respectively. The sensitivity of the devices increases as the frequency is increased up to 20 Hz and then decreases for the frequency range from 20 to 50 Hz as seen from Fig. 6.24b. Beam thinning process optimized to critically control the beam thickness results in shifting the resonance frequency to lower frequency range, thereby improving the overall performance of both the devices in low-frequency domain. A considerably higher sensitivity has been observed for frequency as low as 20 Hz. The shaker at low-frequency testing has a very high amplitude as compared to testing near resonance frequency and therefore both the results shown in Fig. 6.24a, b cannot be compared for maximum device output. From Table 6.4, the experimental and the simulated result are in close agreement with each other. The beam thinning process done using DRIE to reduce the beam thickness of the two-beam and four-beam device resulted in low-frequency operation. As the resonance frequency of the four-beam device (515 Hz) and the two-beam device (466 Hz) are quite close to each other this gives similar low-frequency response for both devices as shown in Fig. 6.24a.

Table 6.2 Sensitivity around resonance frequency

Shaker frequency (Hz)	Two-beam device ($mV/m/s^2$)	Four-beam device ($mV/m/s^2$)
50	1.1250	0.84607
100	1.6076	1.2418
300	2.4066	1.3925
500 (near Resonance)	**1.1392**	**1.0231**
800	0.73915	0.67223
1000	0.58963	0.56588

Table 6.3 Sensitivity for lowest frequency range

Shaker frequency (Hz)	Two-beam device ($mV/m/s^2$)	Four-beam device ($mV/m/s^2$)
10	1.9777	2.2968
20	2.7845	2.3297
50	1.1250	0.84607

Table 6.4 Net sensitivity for two-beam and four-beam device

Sensitivity	Two-beam device (mV/m/s^2)	Four-beam device (mV/m/s^2)
Experimental	2.2784	4.09272
FEM	2.36	4.82

6.5.1 Output Voltage/Power Versus Input Acceleration

The two-beam device is tested using 9100D Portable Shaker Table at Structural Solutions Private Limited, Hyderabad, India, to obtain the output voltage/power as a function of input acceleration. Figure 6.25 shows the device mounted on the 9100D portable shaker.

The amplitude knob seen in Fig. 6.25 is used to vary the acceleration from 0.5 to 5 g, and voltage/power is obtained experimentally at a constant frequency. Figure 6.26 gives the graph of voltage versus input acceleration from 0.5 to 5 g. Voltage at 0.5 g is 0.67 mV and at 5 g is 6.7 mV. Power calculated at resistance of 284 KOhm is also shown in Fig. 6.26.

Fig. 6.17 Low-level CV measurement of device at CSIR-CEERI

(a) (b)

Fig. 6.18 **a** Device mounted on PCB package. **b** Device package sealed for testing

6.5.2 Output Voltage Versus Frequency

Output voltage from the frequency range from 10 to 800 Hz is obtained experimentally using 9100D Portable Shaker. The frequency from 9100D shaker is varied from 10 to 800 Hz at input acceleration of 0.5 g as shown in Fig. 6.27. Similarly, the frequency variation is done for acceleration of 1, 2 and 3 g and output potential is obtained. Figure 6.28 shows the graph between the output voltage and frequency at different acceleration ranging from 0.5 g to 3 g. The device output increases near 500 Hz, i.e., near the resonance frequency of the device. The device exhibits a large operation frequency range from 10 to 800 Hz which can be seen from Fig. 6.28.

Voltage generated at different acceleration ranging from 0.5 to 5 g is measured and compared for two-beam and four-beam devices. Electric potential generated by the two-beam device at 1 g is 1.34 mV and for the four-beam device is 1.0231 mV. As acceleration increases, voltage generated by both devices increases, at 5 g gives a voltage of 6.7 mV for two-beam device and 5.1155 mV for four-beam device as seen from Fig. 6.29.

Table 6.5 gives comparison of the current work with the published work. Two-beam device gives lowest resonance frequency reported till date and good sensitivity around 500 Hz. The four-beam device is firstly reported in the current work and gives better results than two-beam device.

From the above characterization results, it is concluded that both two-beam and four-beam devices generate a good amount of power or voltage which can be utilized in a good way through the design and fabrication of an array of devices integrated in parallel and series configurations to enhance voltage as well as total power to drive an MSN or WSN.

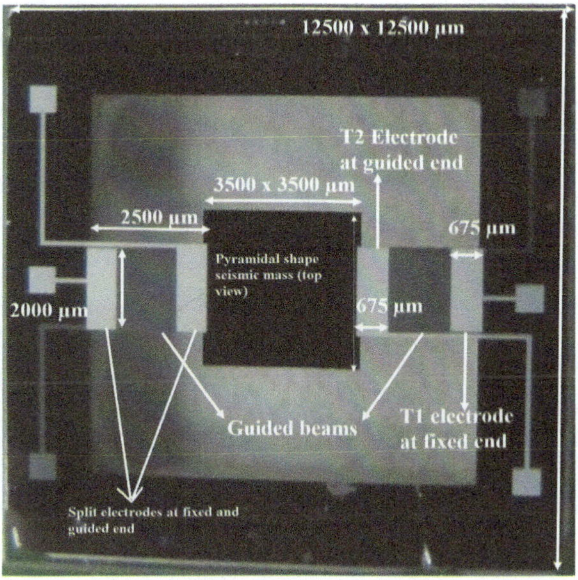

Fig. 6.19 Snapshot of fabricated guided two-beam energy harvester device

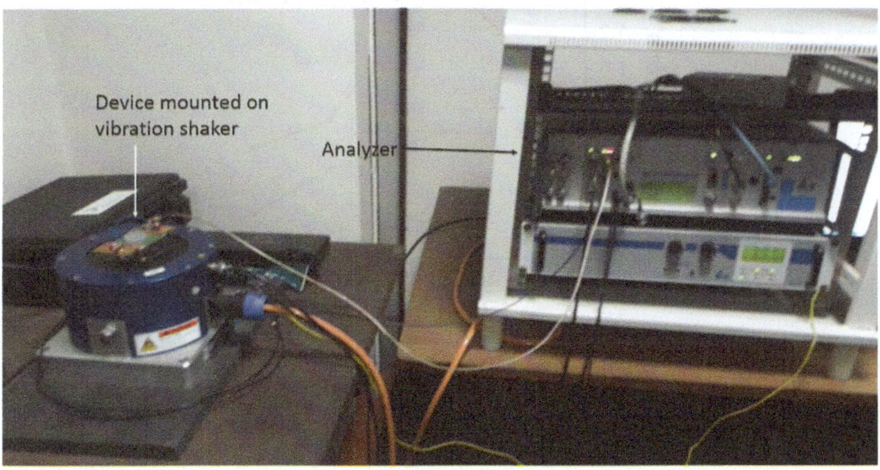

Fig. 6.20 Device mounted on SPEKTRA SE-10 vibration exciter setup at Acoustics and Vibration standards laboratory, National Physical Laboratory, Delhi, India

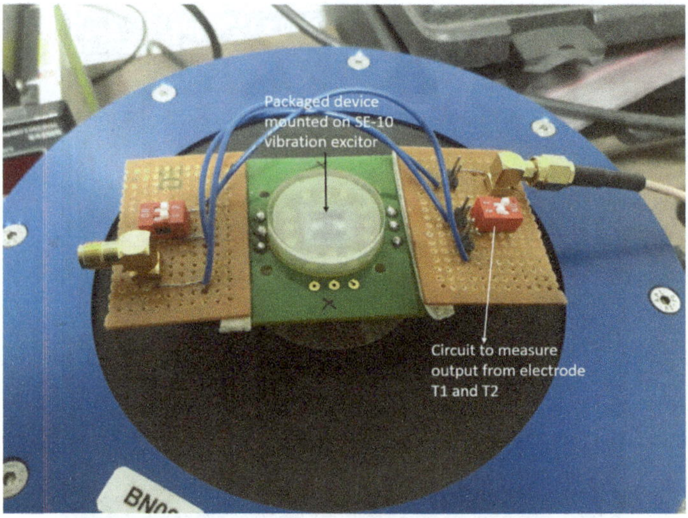

Fig. 6.21 Extended view of the device mounted on SE-10 vibration exciter

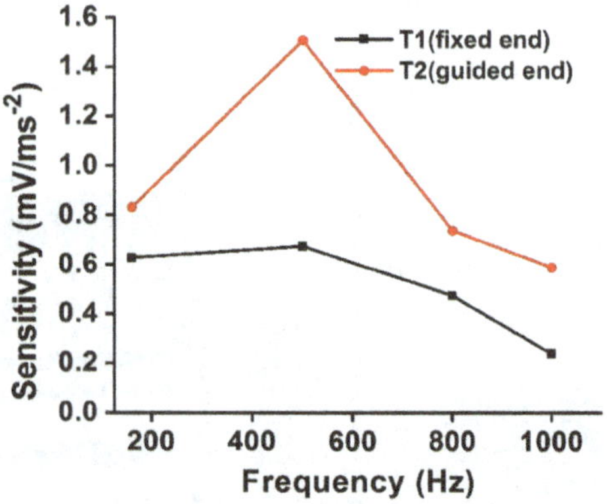

Fig. 6.22 **a** Device Sensitivity measured at T1 and T2 electrodes from 160 to 1000 Hz range

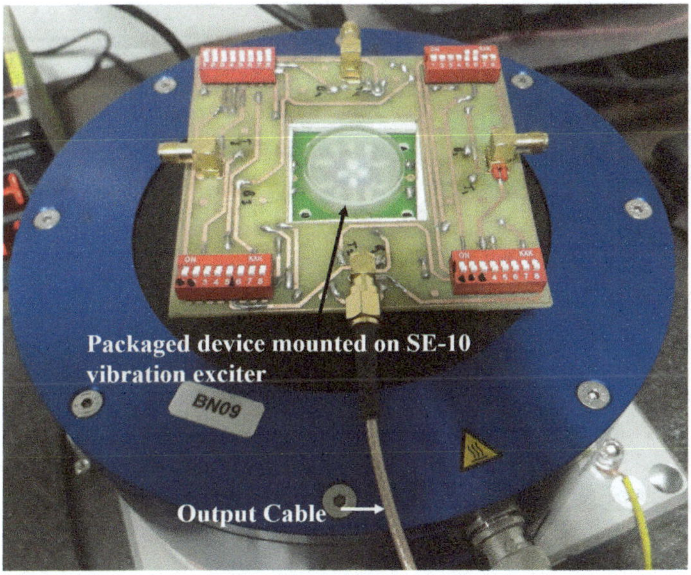

Fig. 6.23 Guided four-beam device mounted on SE-10 vibration exciter

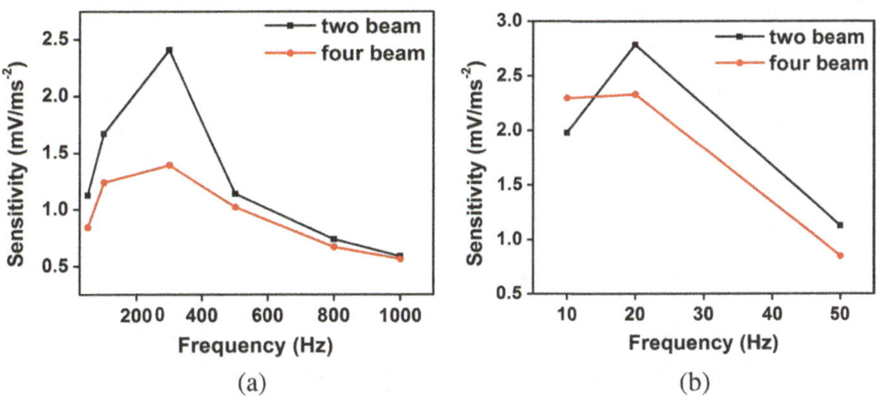

Fig. 6.24 Sensitivity comparison of guided two-beam and four-beam device **a** Lowest frequency range: from 50 to 1000 Hz. **b** Around resonance: from 10 to 50 Hz. Outputs are across single split electrode

Fig. 6.25 Device mounted on 9100D Portable Shaker

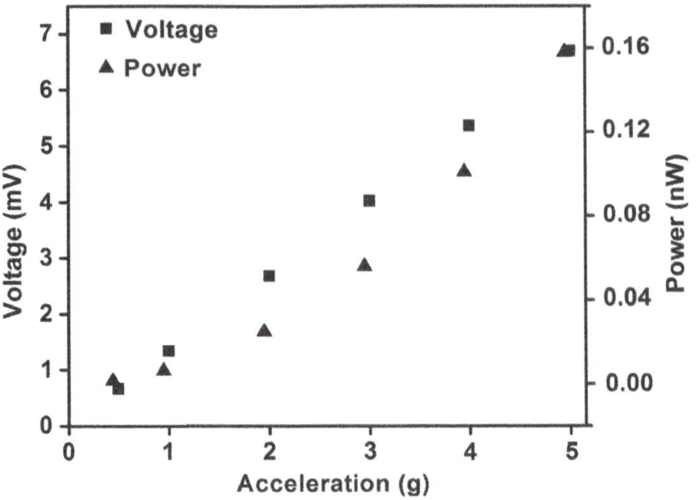

Fig. 6.26 Voltage/power versus acceleration amplitude from 0.5 to 5 g

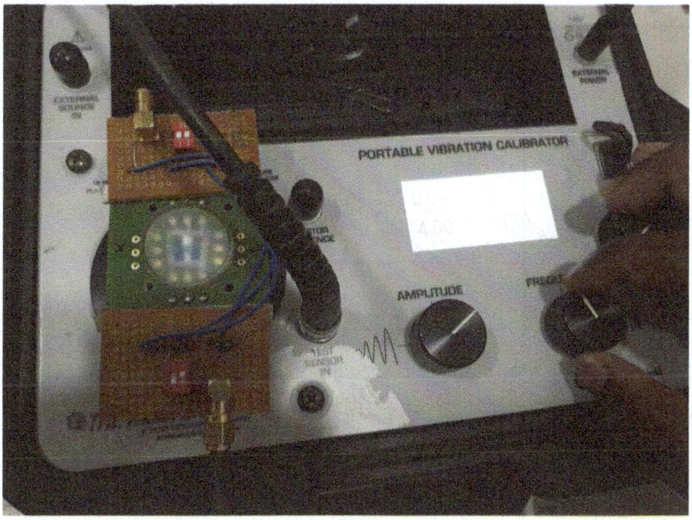

Fig. 6.27 Output voltage versus frequency test using 9100D portable shaker

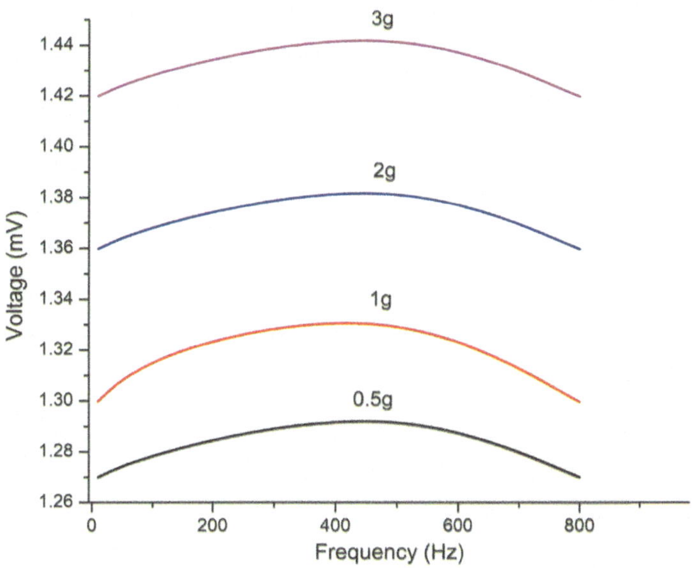

Fig. 6.28 Output voltage and frequency at acceleration ranging from 0.5 to 3 g

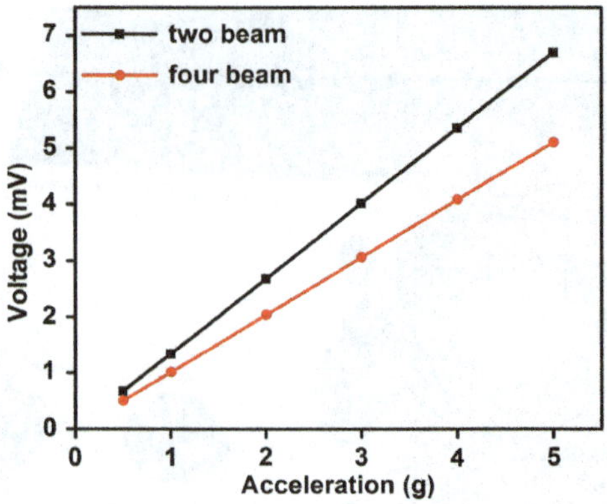

Fig. 6.29 Output voltage and frequency at acceleration ranging from 0.5 to 5 g

Table 6.5 Comparison with published work

S. no:	Marzencki et al. (2009)	Wang et al. (2012)	Fabricated device	Remarks
Resonance frequency	1800 Hz	694 Hz	466/515 Hz	Lower frequency range is better
Output voltage	48.97 mV at 1800 Hz	Not reported	44.64/80.2 mV at 500 Hz	Voltage output is double of the net output, as each beam consists of two pair of electrodes, split electrodes on every beam (configured in series)
Voltage Sensitivity	18.53 mV/m/s^2	Not reported	2.2784/4.09272 mV/m/s^2	Different voltage sensitivity due to different dimensions of seismic mass

References

Marzencki, M., M. Defosseux, and S. Basrour. 2009. MEMS vibration energy harvesting devices with passive resonance frequency adaptation capability. *Journal of Microelectromechanical Systems* 18(6): 1444–1453

Prasad, M., V. Sahula, V.K. Khanna. 2013. Design and fabrication of Si-diaphragm, ZnO piezoelectric film-based MEMS acoustic sensor using SOI wafers. *IEEE transactions on semiconductor manufacturing* 26(2):233–241

Sharma, A., R. Mukhiya, S.S. Kumar, R. Gopal, R., B.D. Pant. 2015. Dynamic characterization of bulk micromachined accelerometer using laser doppler vibrometer (LDV). *Microsystem Technologies* 21(10):2221–2232

Wang, Z., S. Matova, R. Elfrink, M. Jambunathan, C. De Nooijer, R. van Schaijk, and R.J.M. Vullers. 2012. A piezoelectric vibration harvester based on clamped-guided beams. In *2012 IEEE 25th International Conference on Micro Electro Mechanical Systems (MEMS)*, 1201–1204. IEEE

Chapter 7
Design of MEMS-Based Piezoelectric Energy Harvester for Low-Frequency Applications

Abstract This chapter presents the design and FEM analysis of a P-VEH device suitable for practical application. A cantilever-based P-VEH design has been selected because it has lower resonance frequency and high sensitivity which can harvest optimal power from the ambient vibrations. A detailed analysis has been done using FEM tool COMSOL Multiphysics®. Effect of each design parameters such as length, width and thickness of the beam on the cantilever deflection, stress and electric potential generated have been investigated. Parametric variation study has been done where each parameter is increased individually, and the effect on the device performance is investigated. This study enables to optimize the device parameters and select the dimensions which are best suitable for a specific application.

Keywords FEM · Optimization · P-VEH · Cantilever · Displacement · von Mises · Potential

In this chapter, design of P-VEH is presented which is focused on real-time applications. In Chap. 1, practical applications where energy harvesting is suitable were presented in detail. Adequate power, simple design compatible with MEMS fabrication technology makes piezoelectric vibration energy harvesters a preferred choice. Design of an energy harvester is based on the application for which it has been selected. The vibrations frequency range, amplitude, acceleration and conditions for device operation are the key parameters that have to be considered during device design. Based on the aforesaid parameters, a specific design of an energy harvester device is chosen such that it can generate optimal output under the desired frequency range and acceleration. Practical applications where vibrations have a wide frequency range, in such situations a generator array having an array of cantilevers, are selected. Each designed cantilever resonates to a frequency which matches to an ambient vibration frequency giving optimal output in the wide frequency range. Real-time situations where low-frequency–high acceleration vibrations are present, in such scenarios a guided beam structure is more suitable as it can withstand high-amplitude vibrations and gives optimal output.

S. Saxena et al., *Design and Development of MEMS based Guided Beam Type Piezoelectric Energy Harvester*, Energy Systems in Electrical Engineering,
https://doi.org/10.1007/978-981-16-0606-9_7

Table 7.1 Low-frequency Vibrations

Application domain	Vibration frequency range (Hz)	References
Motor vibrations (car, engines)	30–300	Jha et al. [38] [1976]
Human body vibrations	~400	Koenig et al. [39] [2008]
Chopper vibrations	500–2000	Szefi et al. [40] [2004]

7.1 Low-Frequency Vibration Energy Harvesting

Ambient vibrations in different amplitude and low frequency range up to 2 kHz are freely available in the environment. Vibrations frequency range suitable for different application is shown in Table 7.1. It can be seen that the vibration range available for practical applications varies from few Hz up to 2000 Hz. The vibration frequency is generally low frequency in nature because it is either generated from mechanical vibrations in a motor, a chopper or due to human body vibrations such as heartbeat. It can be observed from the available frequency range for the practical applications, that a low frequency and a high output design is suitable for vibration energy harvesting in such conditions.

Considering low-frequency operations and high output requirement, a cantilever-based P-VEH is a suitable design because it has lower resonance frequency and high sensitivity which can harvest optimal power from the ambient vibrations.

7.2 Design of Cantilever Type P-VEH

In this section, a detailed structural and electrical analysis is presented for a cantilever-based P-VEH. In depth analysis is done using FEM tool COMSOL Multiphysics®. Effect of each design parameter such as length, width and thickness of the beam on the cantilever deflection, stress and electric potential generated is investigated. Parametric variation study is done where each parameter is increased individually, and effect on the device performance is investigated. This study enables to optimize the device parameters and select the dimensions which are best suitable for the specific application. Displacement, stress and electric potential are the key parameters which are considered during designing of an energy harvester device. In the subsequent sections, effect of each design parameter on the device displacement, stress and electric potential is presented.

7.3 Effect of Length and Thickness of Piezoelectric Layer on Cantilever Type P-VEH

The effect of length and thickness of the piezoelectric layer on the device displacement, stress and electric potential is investigated and presented in the subsequent sections.

7.3.1 Displacement

Device displacement is the first parameter considered for designing any energy harvester device. The maximum displacement of the cantilever beam will result in the maximum stress in the beam which will result in generation of charge in the piezoelectric layer. Maximum displacement also defines the depth of the cavity required for the device to avoid collisions with the encapsulation layer. In this investigation, length (l_{piezo}) and thickness (t_{piezo}) of the piezoelectric layer are varied and device displacement is obtained. Parametric study is done by varying the length of the piezoelectric layer from 50 to 200 μm having an increment of 50 μm. Three values of the piezoelectric layer thickness, i.e., 2, 5 and 10 μm, are selected. The length of the silicon beam L_{canti} is 200 μm, and beam thickness (T_{canti}) is selected as 10 μm. Width of the cantilever beam L_w is 50 μm, and a seismic mass is placed at the bottom side having length 50 μm, width 50 μm and depth 30 μm.

The FEM results for the cantilever displacement are obtained for the piezoelectric layer thickness (t_{piezo}) as 2 μm is shown below (Figs. 7.1, 7.2, 7.3, 7.4, 7.5, 7.6, 7.7 and 7.8).

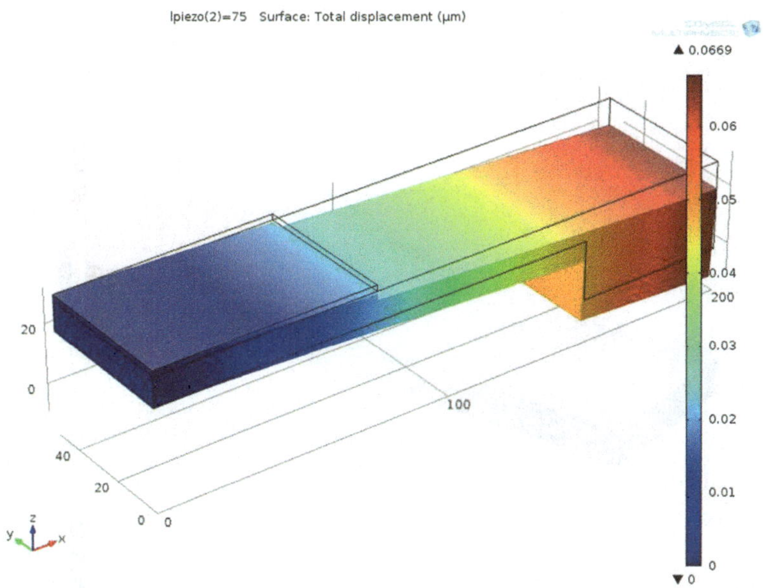

Fig. 7.1 Displacement when $l_{piezo} = 50$ μm

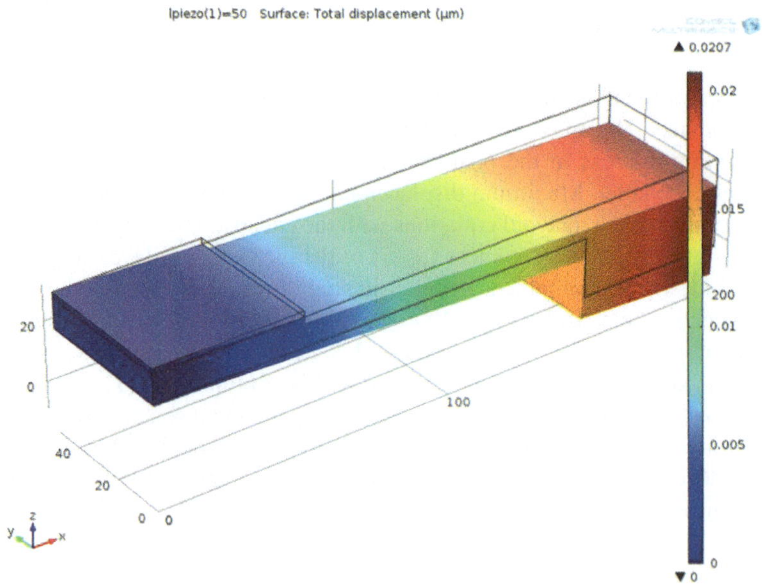

Fig. 7.2 Displacement when $l_{piezo} = 75\ \mu m$

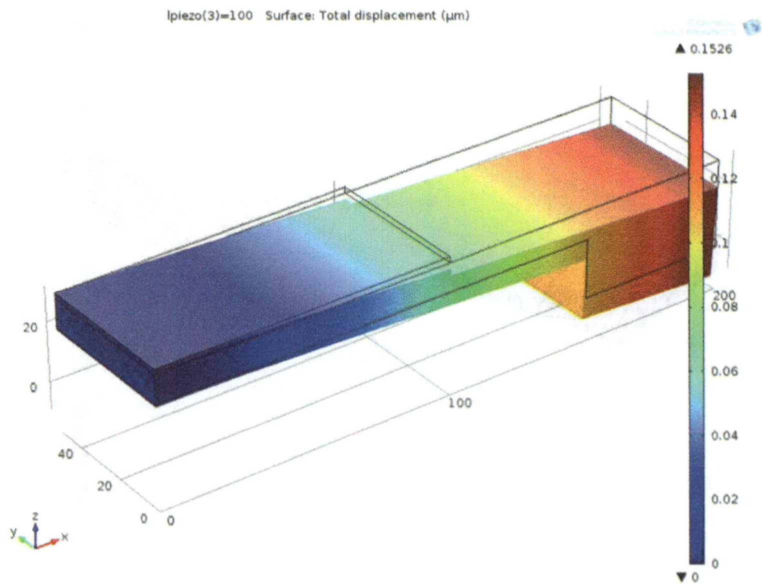

Fig. 7.3 Displacement when $l_{piezo} = 100\ \mu m$

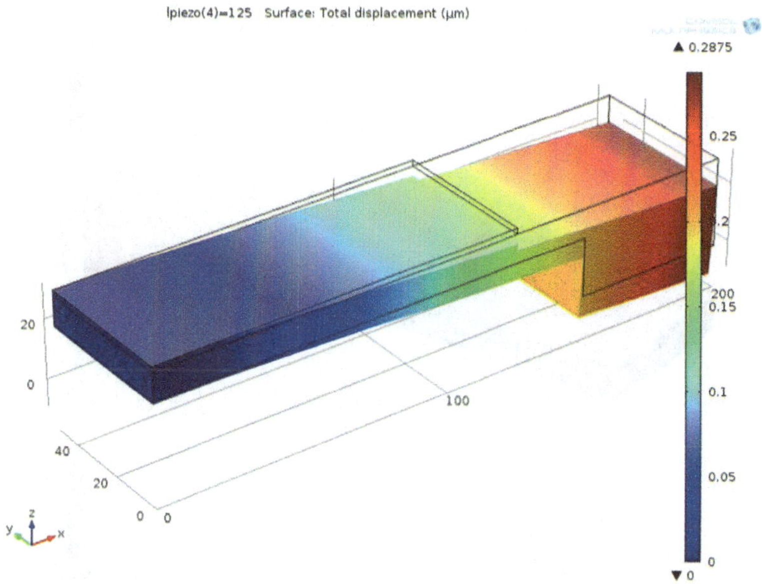

Fig. 7.4 Displacement when $l_{piezo} = 125\ \mu m$

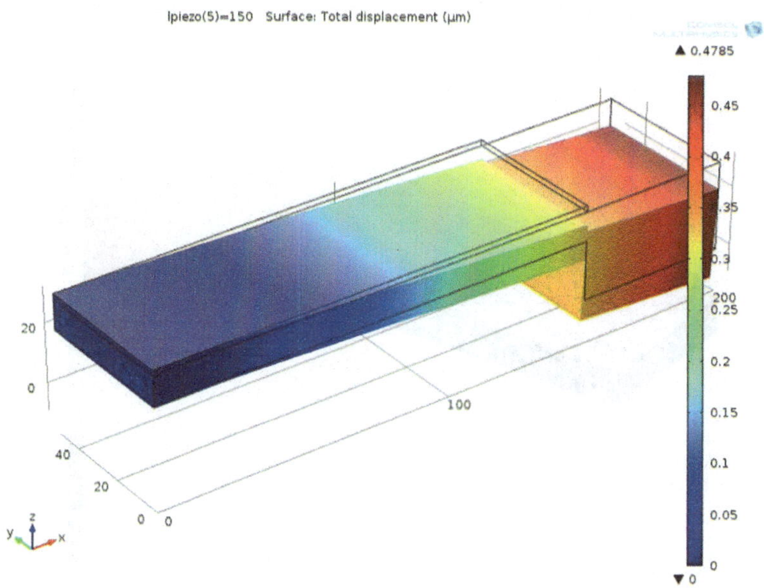

Fig. 7.5 Displacement when $l_{piezo} = 150\ \mu m$

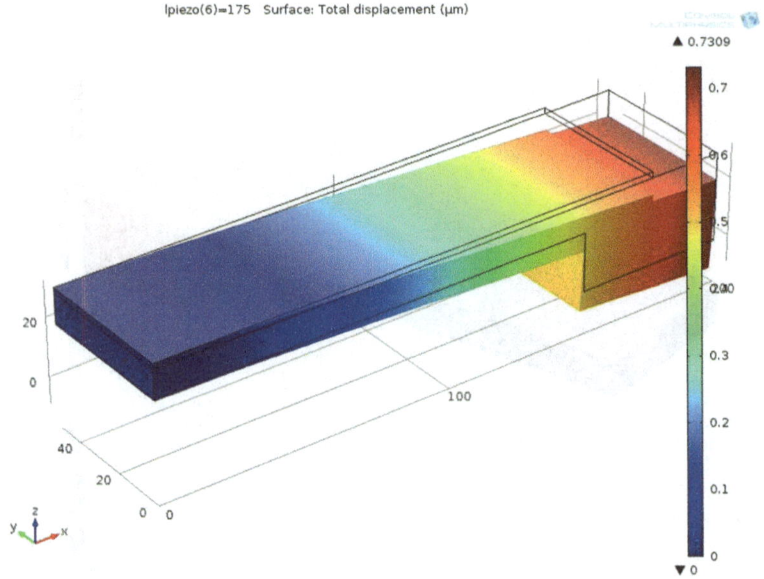

Fig. 7.6 Displacement when $l_{\text{piezo}} = 175\ \mu\text{m}$

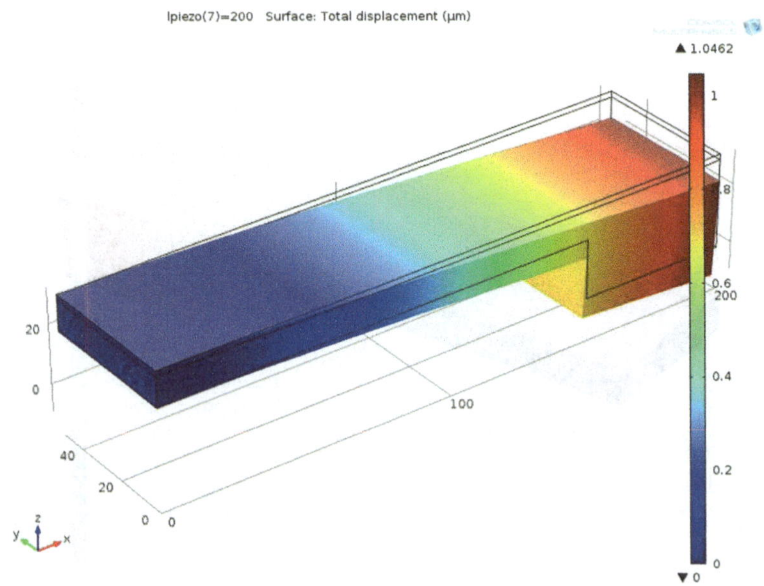

Fig. 7.7 Displacement when $l_{\text{piezo}} = 200\ \mu\text{m}$

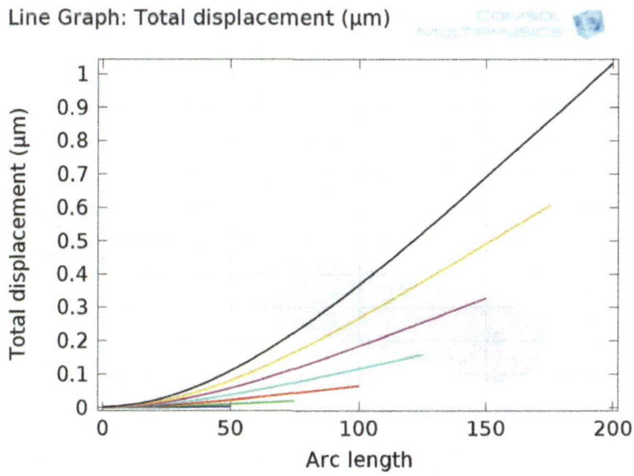

Fig. 7.8 Displacement versus piezoelectric layer length

It can be seen from the above analysis that as the length of the piezoelectric layer increases the displacement also increases. The displacement is minimum at the fixed end and increases while moving toward the free end as shown in Fig. 7.8.

The FEM results for the cantilever displacement are obtained for the piezoelectric layer thickness (t_{piezo}) as 5 μm is shown below (Figs. 7.9, 7.10, 7.11, 7.12, 7.13, 7.14, 7.15 and 7.16).

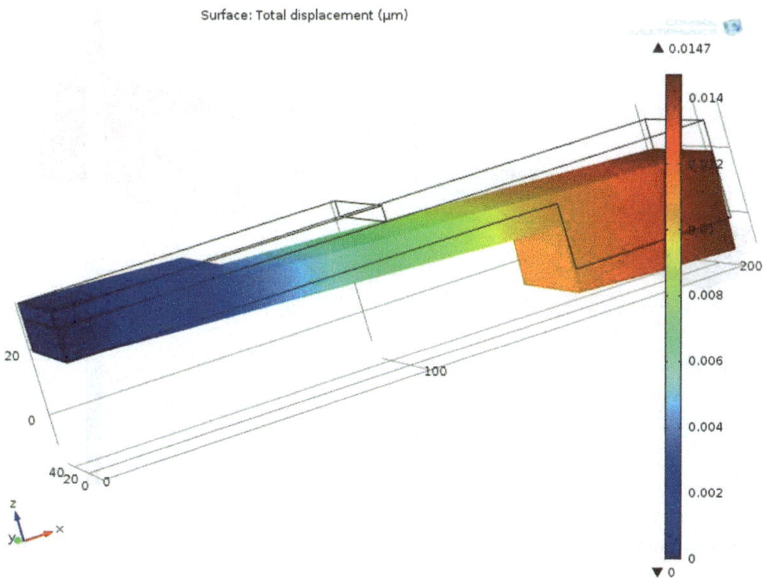

Fig. 7.9 Displacement when $l_{piezo} = 50\ \mu\text{m}$

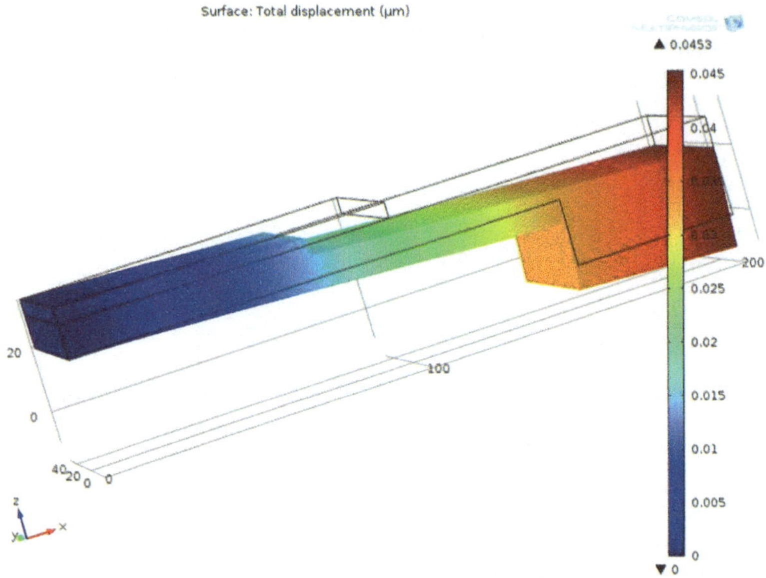

Fig. 7.10 Displacement when $l_{piezo} = 75\ \mu m$

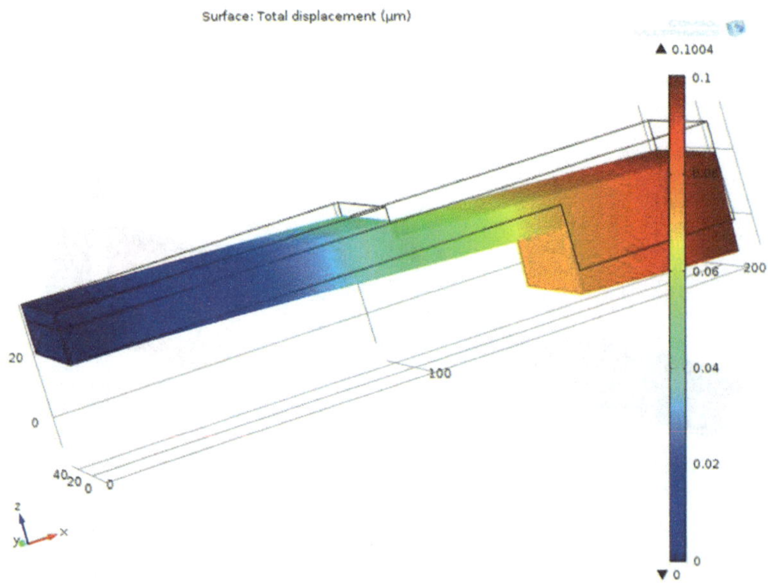

Fig. 7.11 Displacement when $l_{piezo} = 100\ \mu m$

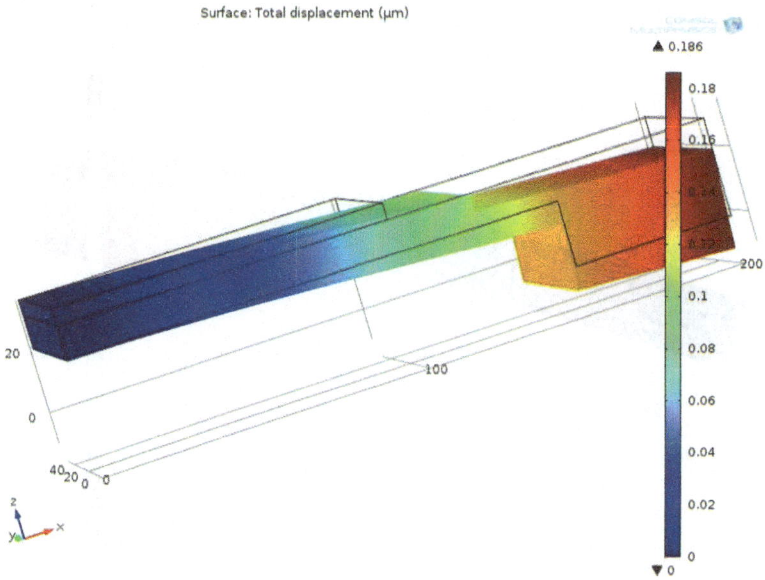

Fig. 7.12 Displacement when $l_{\text{piezo}} = 125\ \mu\text{m}$

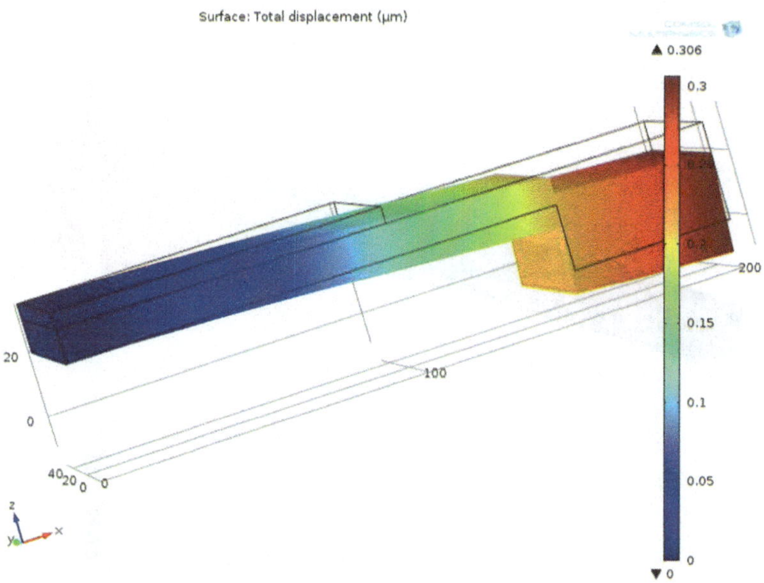

Fig. 7.13 Displacement when $l_{\text{piezo}} = 150\ \mu\text{m}$

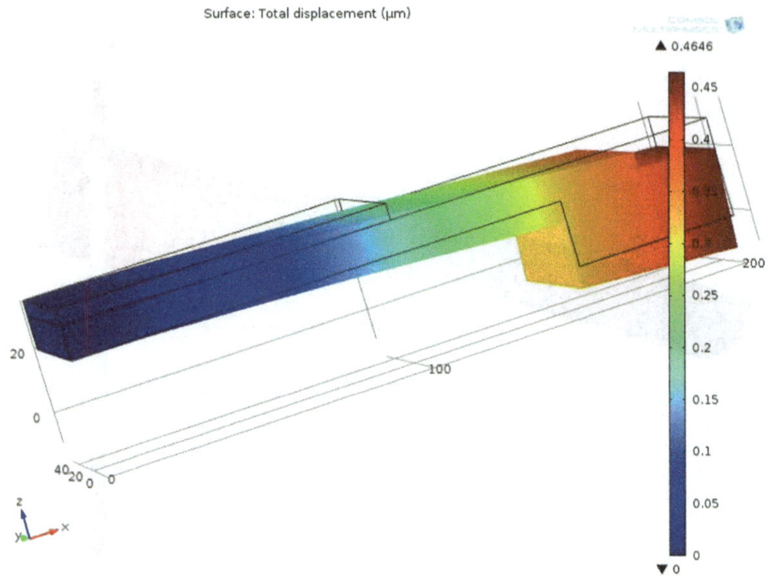

Fig. 7.14 Displacement when $l_{piezo} = 175\ \mu m$

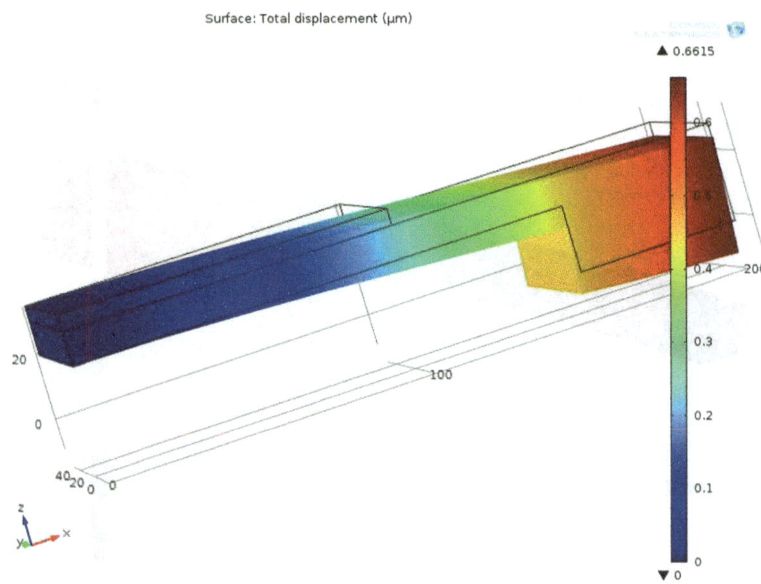

Fig. 7.15 Displacement when $l_{piezo} = 200\ \mu m$

Fig. 7.16 Displacement versus piezoelectric layer length

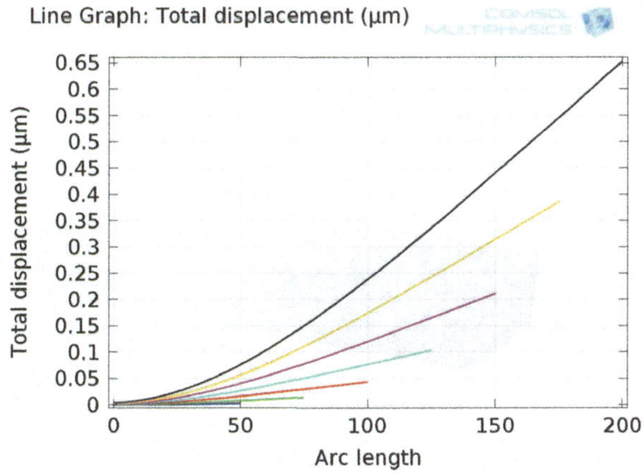

The FEM results for the cantilever displacement are obtained for the piezoelectric layer thickness (t_{piezo}) as 10 µm is shown below (Figs. 7.17, 7.18, 7.19, 7.20, 7.21, 7.22, 7.23 and 7.24).

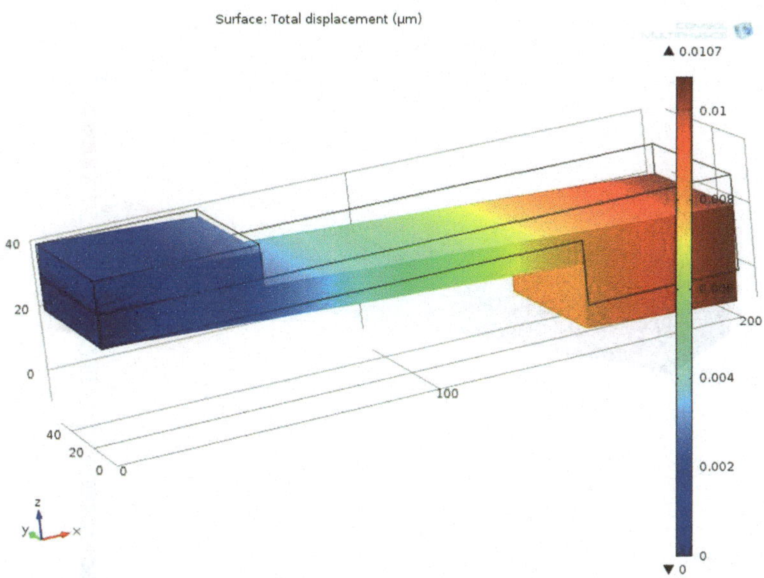

Fig. 7.17 Displacement when $l_{piezo} = 50$ µm

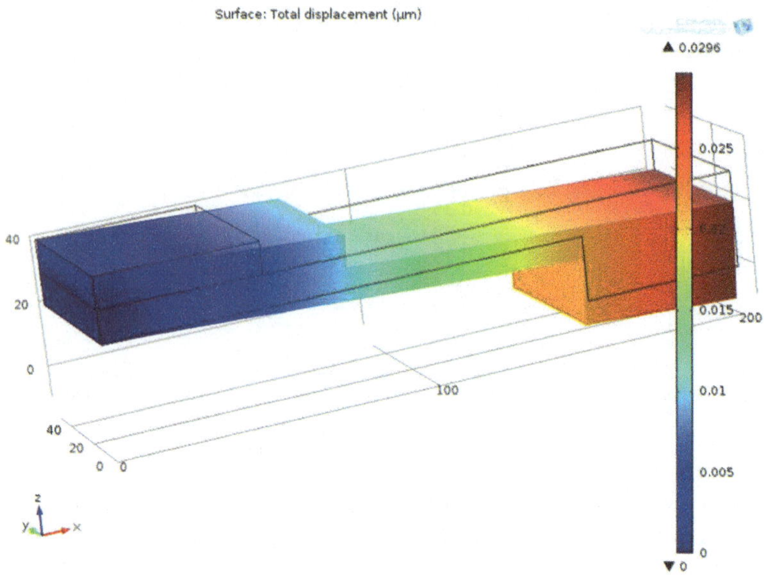

Fig. 7.18 Displacement when $l_{\text{piezo}} = 75\ \mu\text{m}$

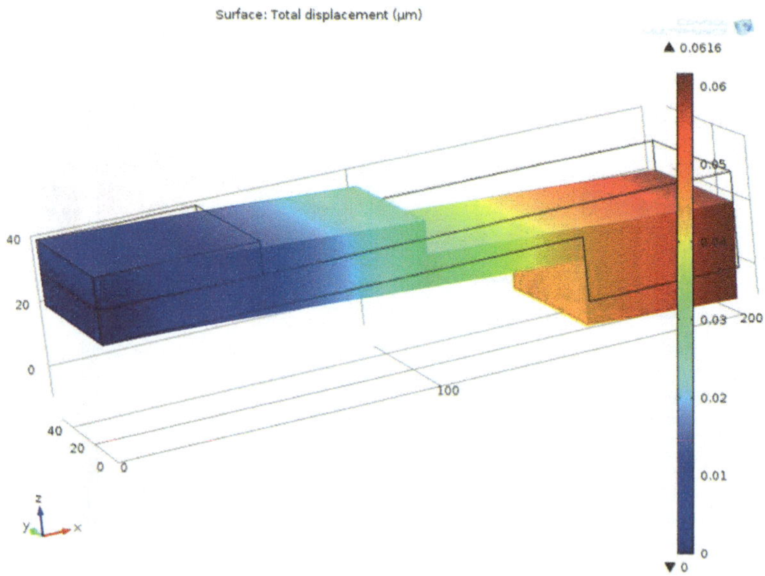

Fig. 7.19 Displacement when $l_{\text{piezo}} = 100\ \mu\text{m}$

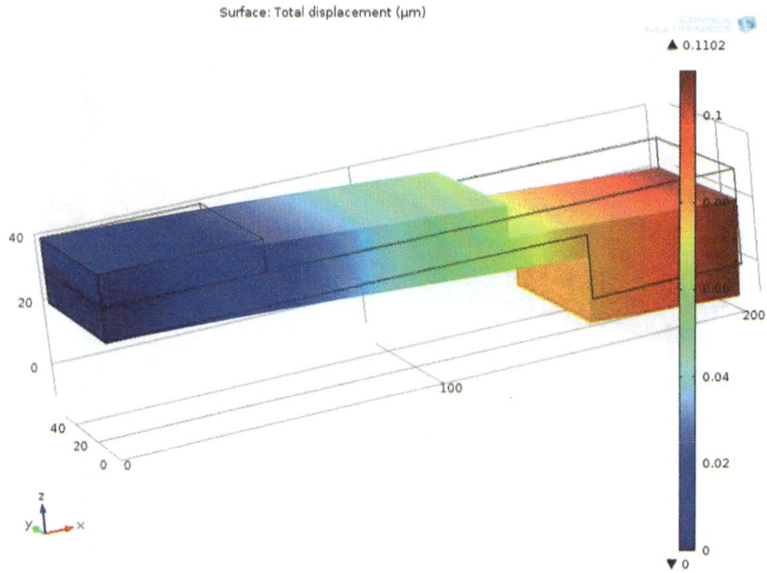

Fig. 7.20 Displacement when $l_{piezo} = 125\ \mu m$

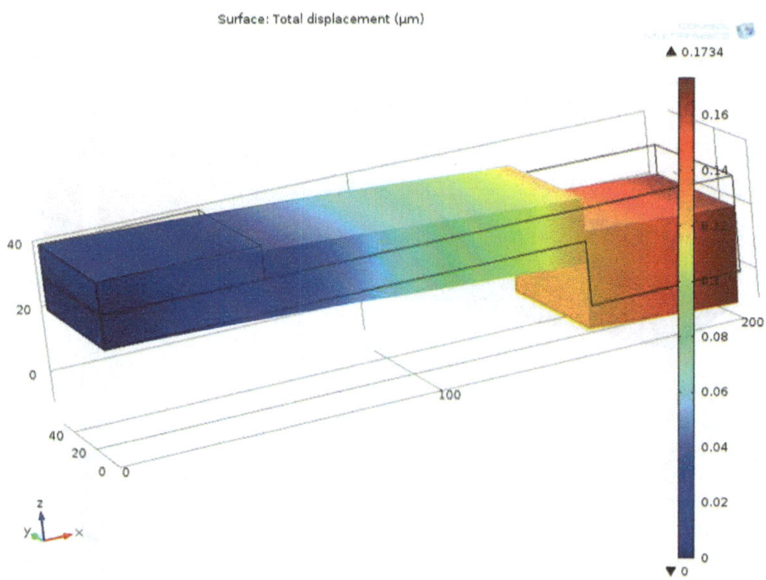

Fig. 7.21 Displacement when $l_{piezo} = 150\ \mu m$

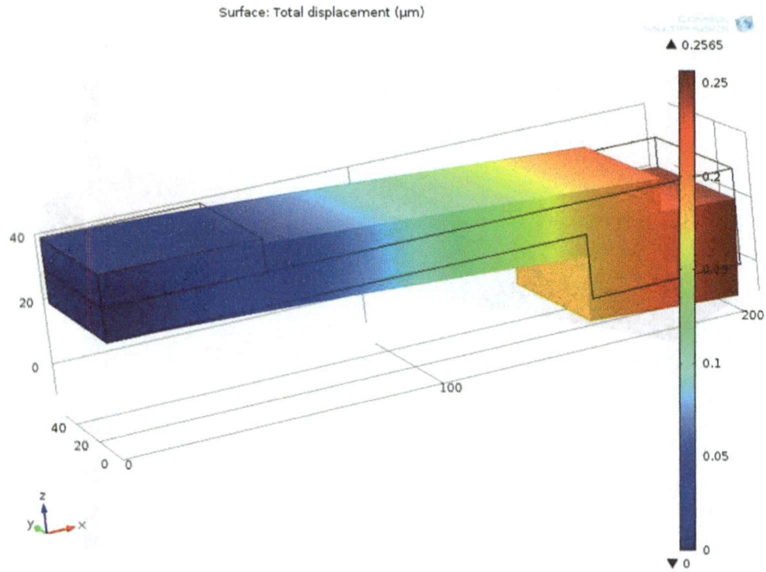

Fig. 7.22 Displacement when $l_{piezo} = 175\ \mu m$

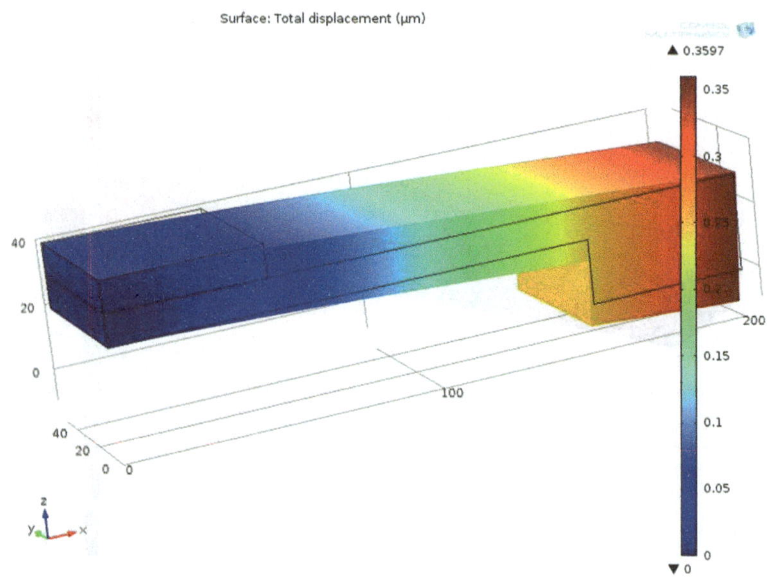

Fig. 7.23 Displacement when $l_{piezo} = 200\ \mu m$

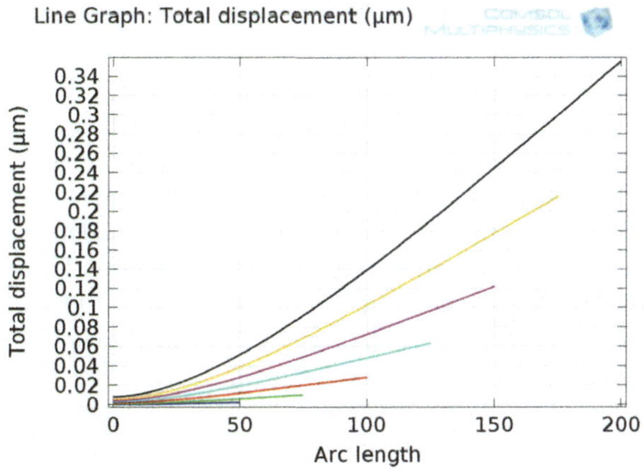

Fig. 7.24 Displacement versus piezoelectric layer length

It can be seen from the above analysis that the length of the piezoelectric layer has an effect on the cantilever displacement. From Table 7.2, at a piezoelectric layer thickness t_{piezo} of 2 µm, displacement of 0.020 µm is obtained for a piezoelectric layer length of 50 µm and at a length of 200 µm, the displacement increases to 1.046 µm. Increasing the thickness of the piezoelectric layer to 5 and 10 µm results in a reduction of the cantilever displacement. This reduction is due to the fact that as the thickness of the layer is increased the stiffness in the beam increases which reduces the displacement of the beam. This analysis is important because while designing the harvester the device displacement must be known which will generate the stress along the cantilever beam which results in electric potential generation. Therefore, the thickness of the piezoelectric layer must be chosen carefully for higher electric potential and beam displacement.

Table 7.2 Displacement versus length for different thickness of piezoelectric layer

S. No.	l_{piezo} (µm)	Displacement (µm)		
		$t_{piezo} = 2$ µm	$t_{piezo} = 5$ µm	$t_{piezo} = 10$ µm
1	50	0.020	0.014	0.010
2	75	0.066	0.045	0.029
3	100	0.152	0.100	0.061
4	125	0.287	0.186	0.110
5	150	0.478	0.306	0.173
6	175	0.730	0.464	0.256
7	200	1.046	0.661	0.359

7.3.2 von Mises Stress

Stress is generated along the cantilever beam due to displacement under external vibration. Along X-axis stress is generated which is also exerted by the piezoelectric layer on the cantilever beam. A highly c-axis oriented piezoelectric layer is deposited which operates on d31 mode which results in generation of charge along the length of the beam. In this investigation, length (l_{piezo}) and thickness (t_{piezo}) of the piezoelectric layer are varied and von Mises stress is obtained. Parametric study is done by varying the length of the piezoelectric layer from 50 to 200 μm having an increment of 50 μm. Three values of the piezoelectric layer thickness, i.e., 2, 5 and 10 μm, are selected. The length of the silicon beam (L_{canti}) is 200 μm, and beam thickness (T_{canti}) is selected as 10 μm. Width of the cantilever beam L_w is 50 μm, and a seismic mass is placed at the bottom side having length 50 μm, width 50 μm and depth 30 μm.

The FEM results for the von Mises stress are obtained for the piezoelectric layer thickness (t_{piezo}) as 2 μm is shown below (Figs. 7.25, 7.26, 7.27, 7.28, 7.29, 7.30, 7.31 and 7.32).

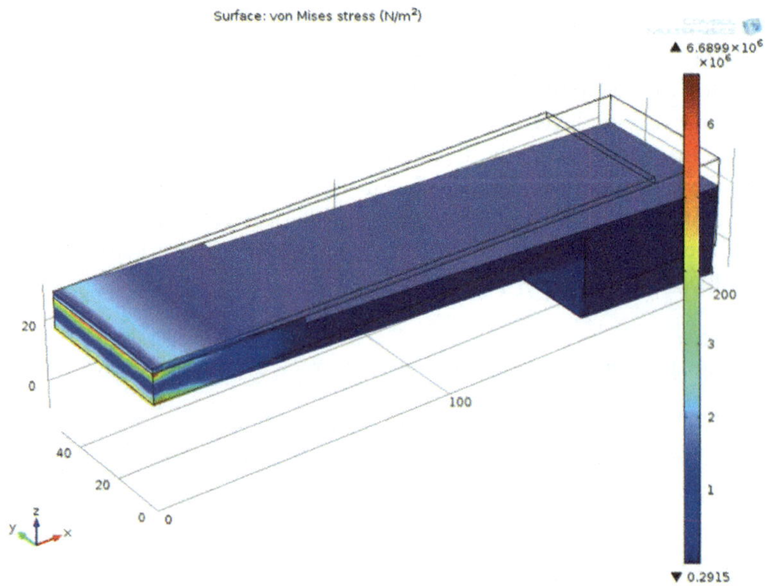

Fig. 7.25 von Mises stress when $l_{piezo} = 50$ μm

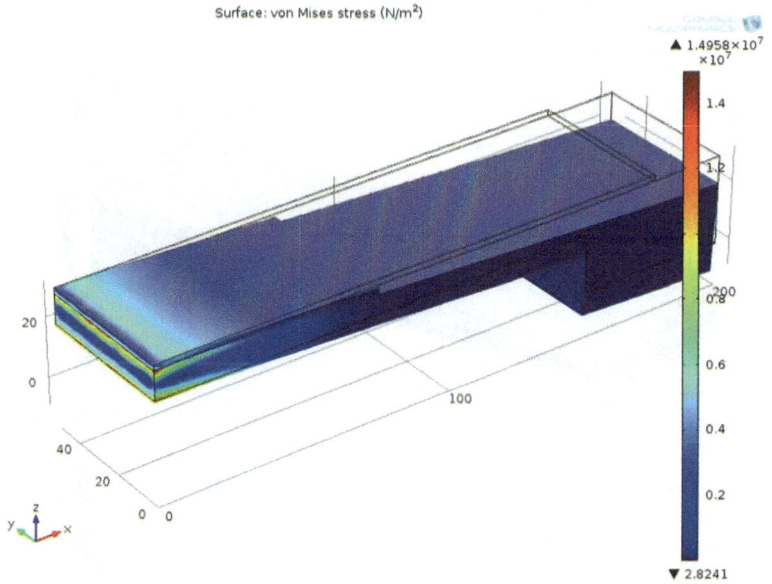

Fig. 7.26 von Mises stress when $l_{\text{piezo}} = 75\ \mu\text{m}$

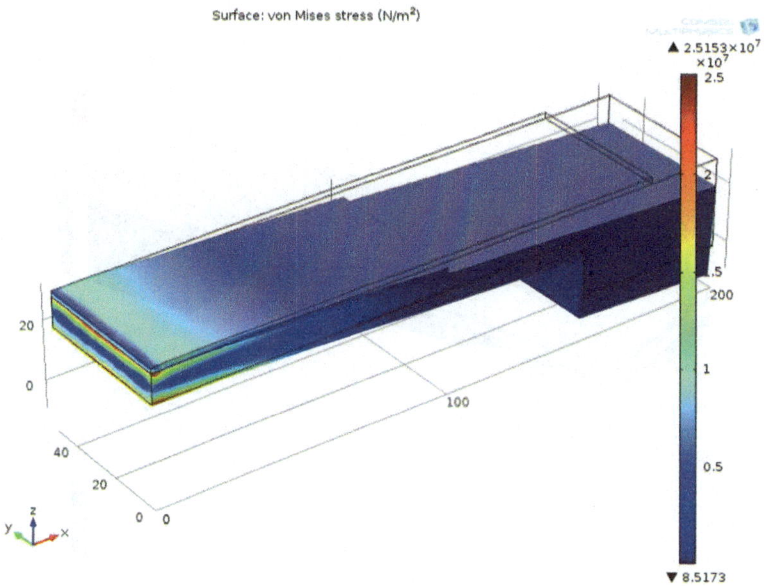

Fig. 7.27 von Mises stress when $l_{\text{piezo}} = 100\ \mu\text{m}$

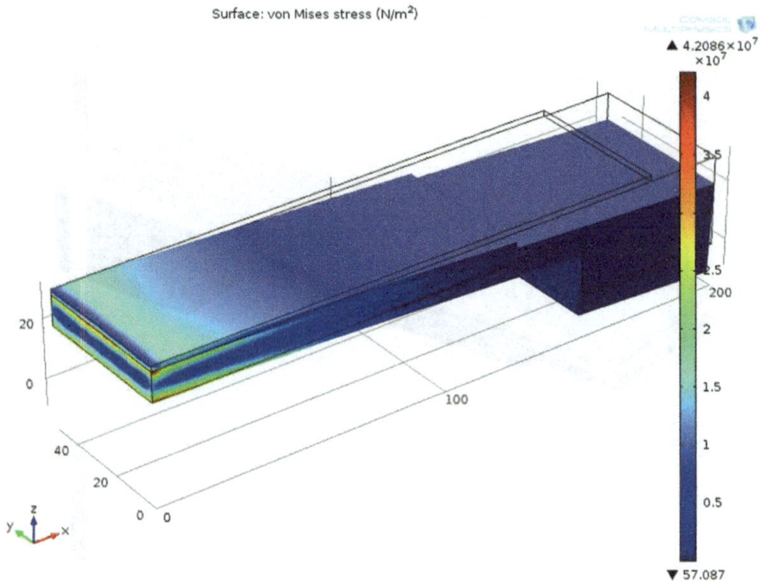

Fig. 7.28 von Mises stress when $l_{piezo} = 125\ \mu m$

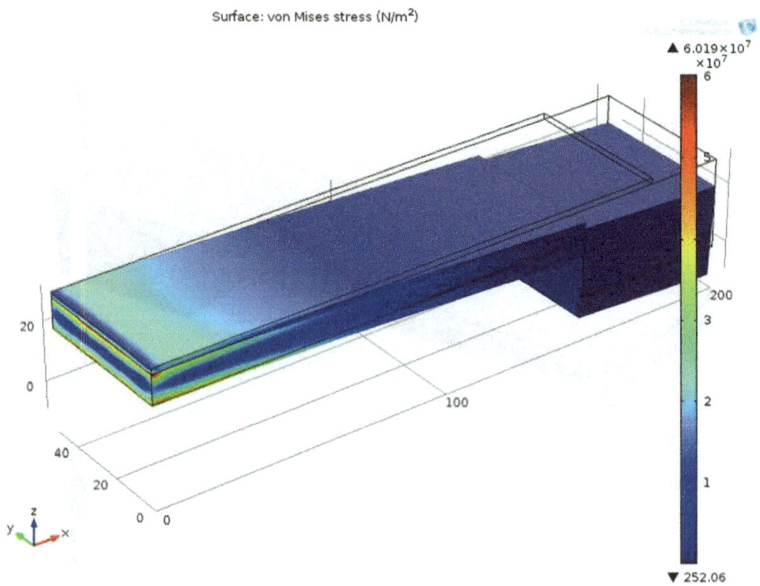

Fig. 7.29 von Mises stress when $l_{piezo} = 150\ \mu m$

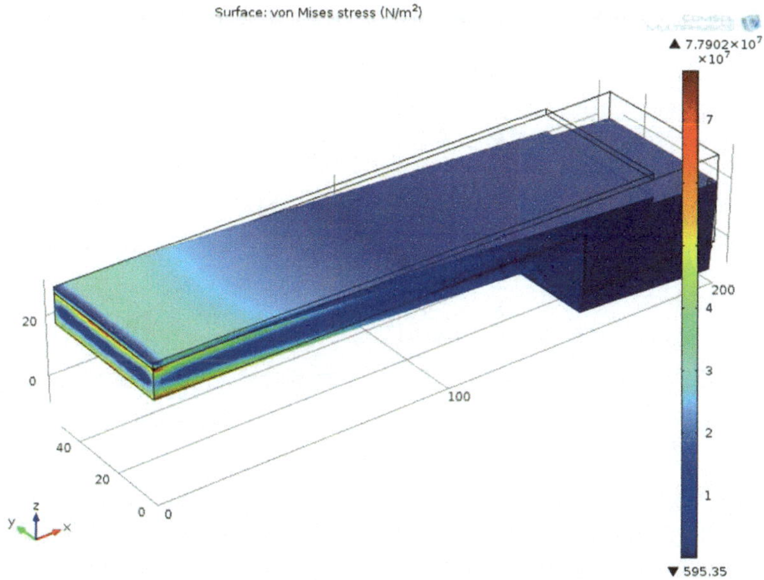

Fig. 7.30 von Mises stress when $l_{\text{piezo}} = 175\ \mu\text{m}$

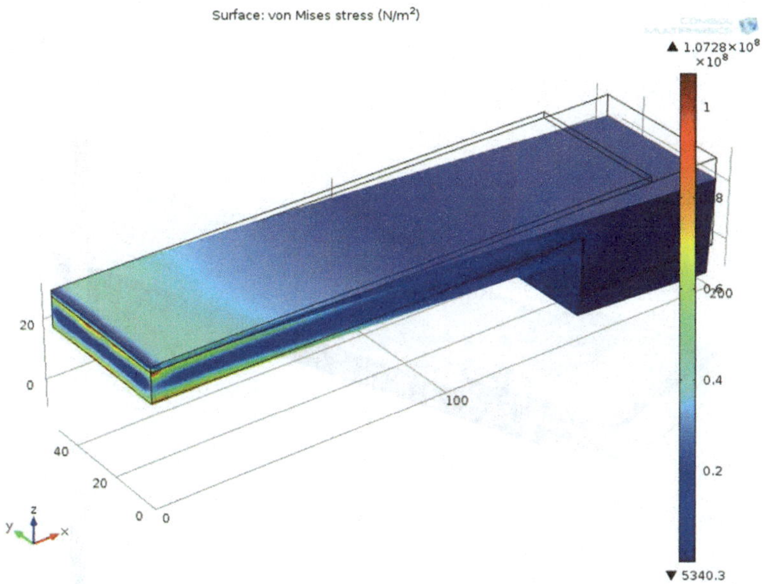

Fig. 7.31 von Mises stress when $l_{\text{piezo}} = 200\ \mu\text{m}$

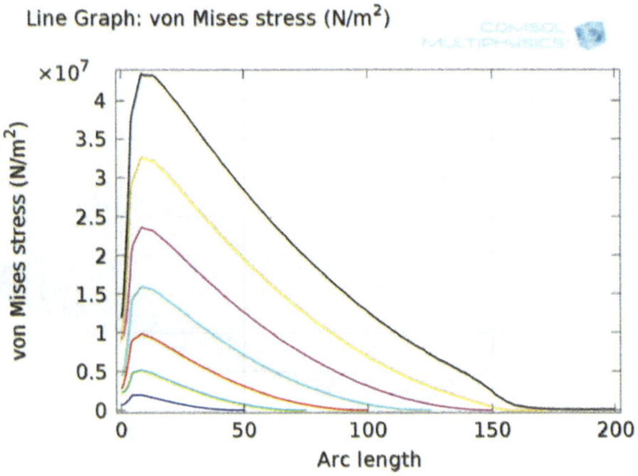

Fig. 7.32 von Mises stress versus piezoelectric layer length

It can be seen from the above analysis that the stress is maximum at the fixed end which decreases gradually while moving toward free end. As the length of the piezoelectric layer increases, the stress on the cantilever beam increases.

The FEM results for the von Mises stress are obtained for the piezoelectric layer thickness (t_{piezo}) as 5 μm is shown below (Figs. 7.33, 7.34, 7.35, 7.36, 7.37, 7.38, 7.39 and 7.40).

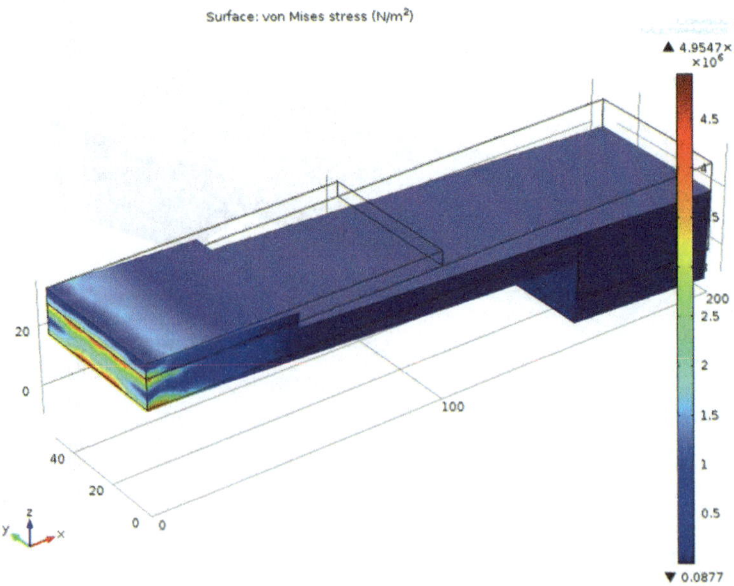

Fig. 7.33 von Mises stress when $l_{piezo} = 50$ μm

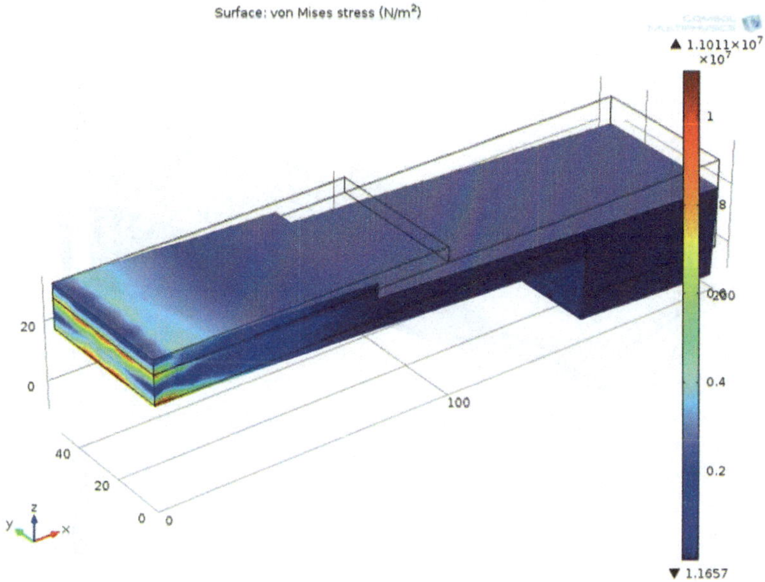

Fig. 7.34 von Mises stress when $l_{\text{piezo}} = 75\ \mu\text{m}$

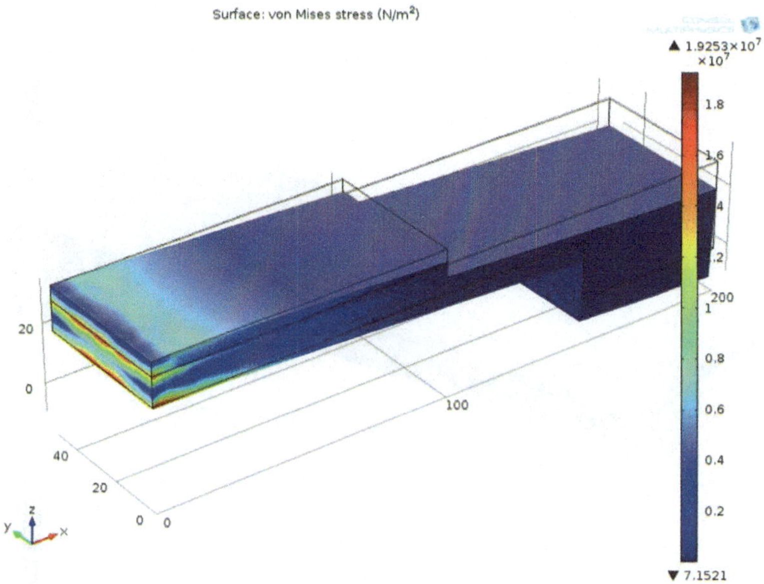

Fig. 7.35 von Mises stress when $l_{\text{piezo}} = 100\ \mu\text{m}$

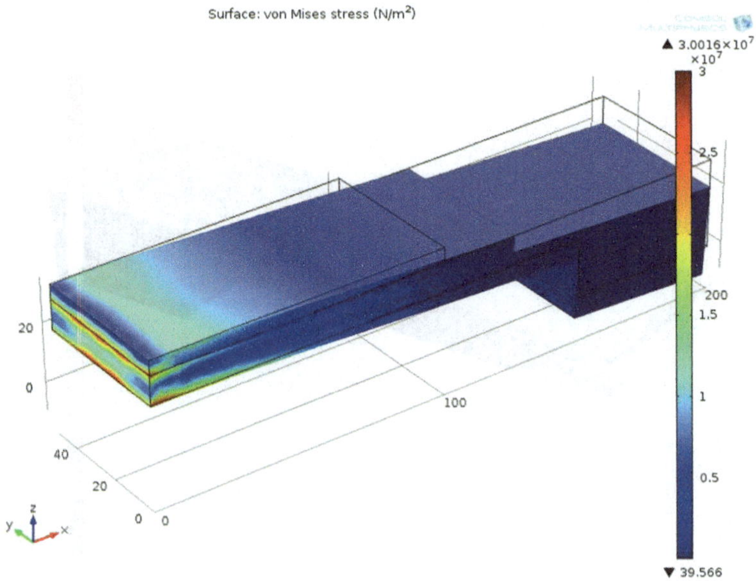

Fig. 7.36 von Mises stress when $l_{\text{piezo}} = 125\ \mu\text{m}$

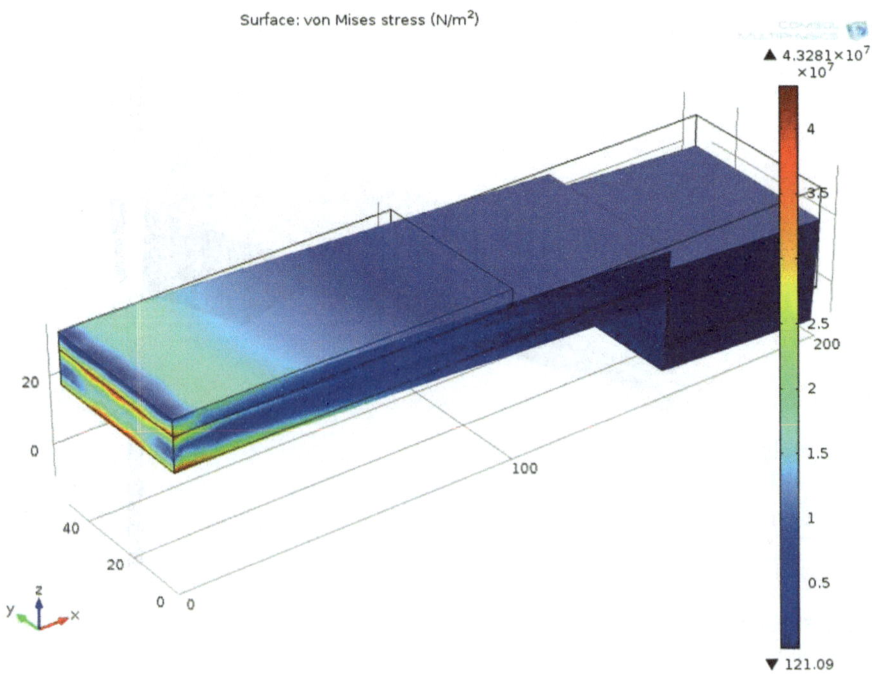

Fig. 7.37 von Mises stress when $l_{\text{piezo}} = 150\ \mu\text{m}$

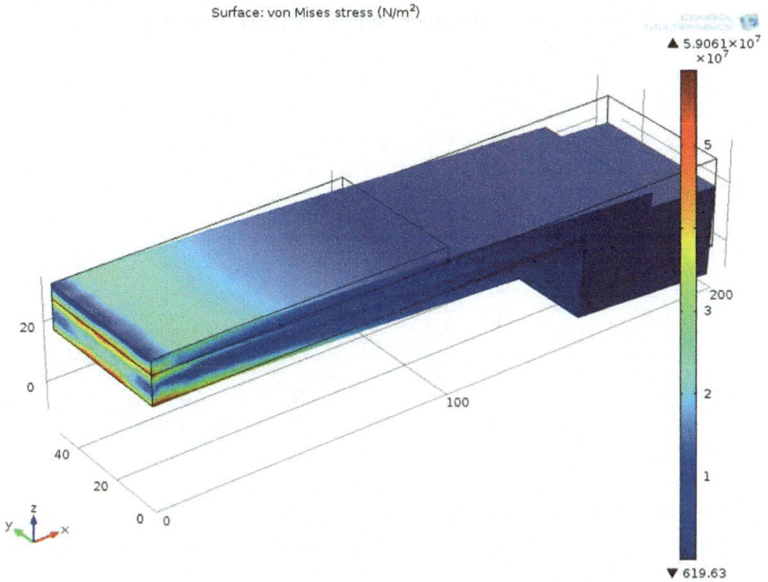

Fig. 7.38 von Mises stress when $l_{\text{piezo}} = 175\ \mu\text{m}$

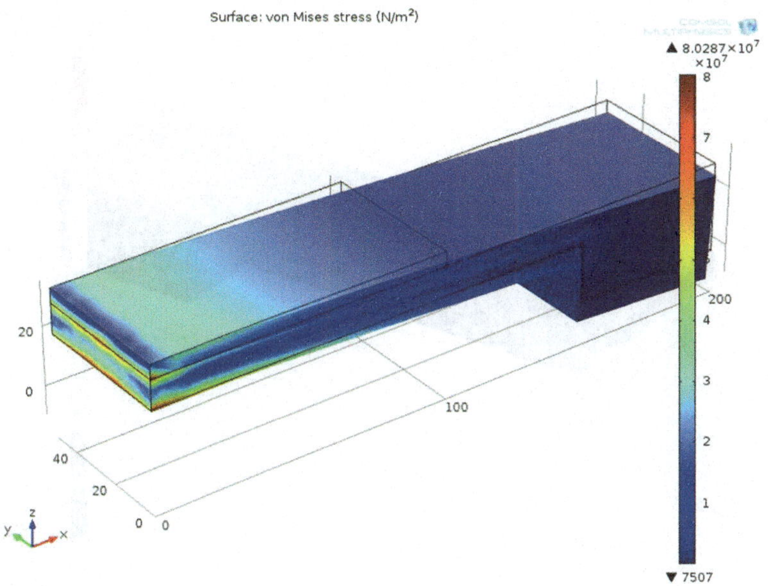

Fig. 7.39 von Mises stress when $l_{\text{piezo}} = 200\ \mu\text{m}$

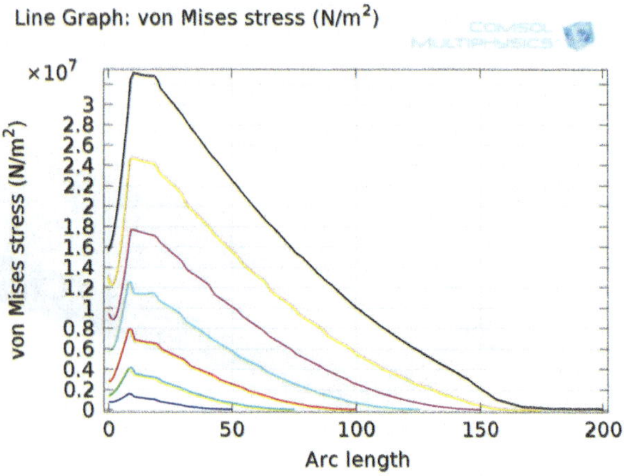

Fig. 7.40 von Mises stress versus piezoelectric layer length

The FEM results for the von Mises stress are obtained for the piezoelectric layer thickness (t_{piezo}) as 10 μm is shown below (Figs. 7.41, 7.42, 7.43, 7.44, 7.45, 7.46, 7.47 and 7.48).

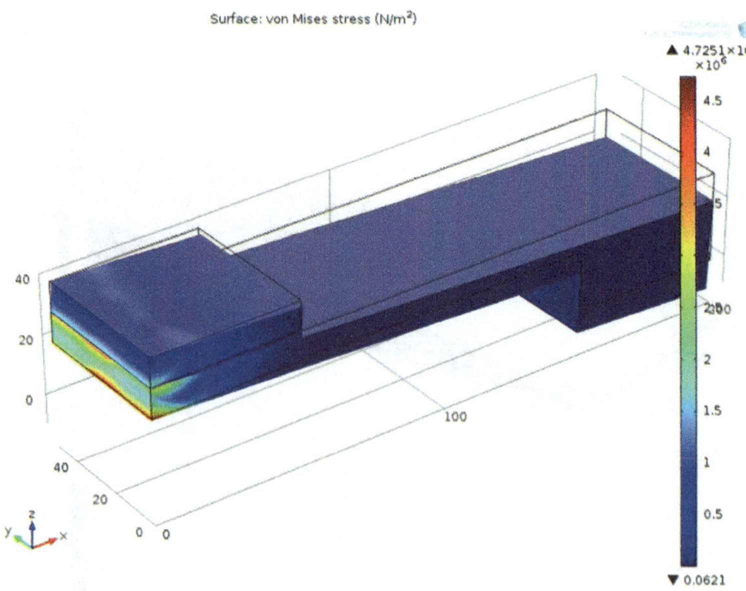

Fig. 7.41 von Mises stress when $l_{\text{piezo}} = 50$ μm

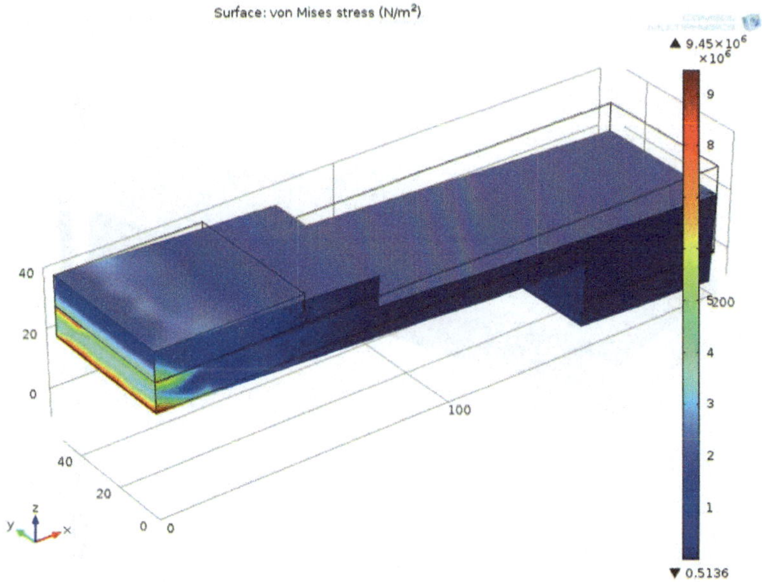

Fig. 7.42 von Mises stress when $l_{\text{piezo}} = 75\ \mu\text{m}$

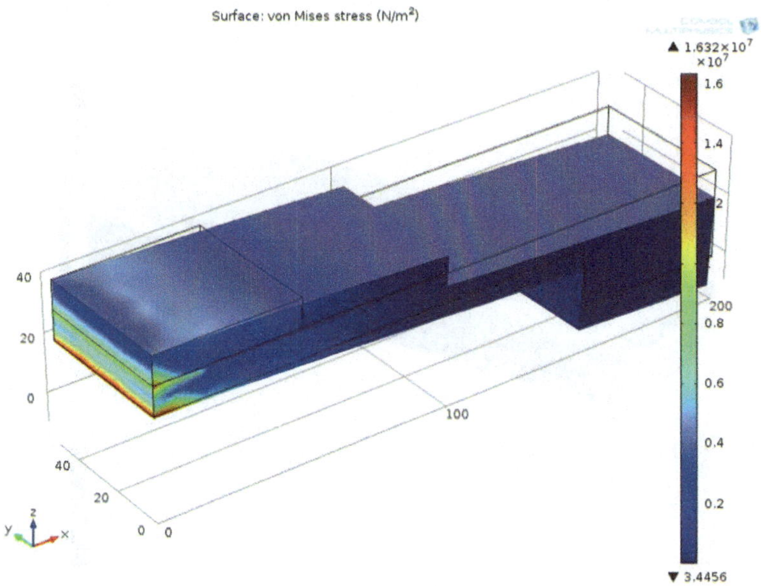

Fig. 7.43 von Mises stress when $l_{\text{piezo}} = 100\ \mu\text{m}$

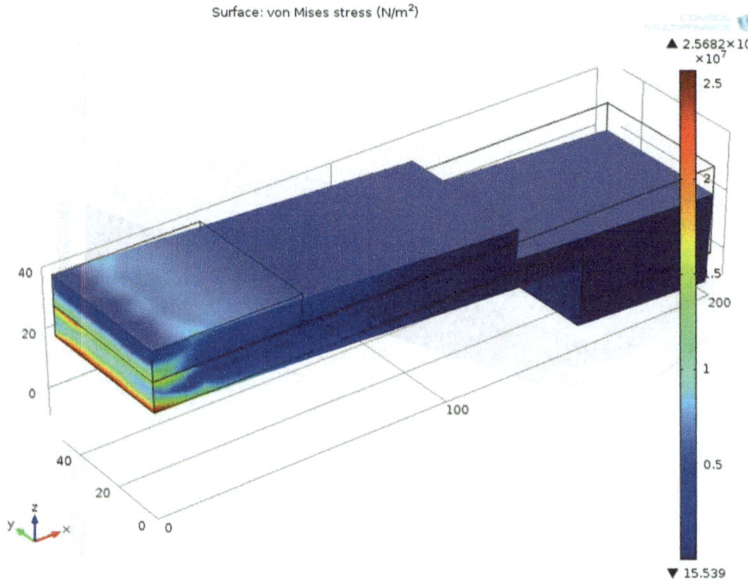

Fig. 7.44 von Mises stress when $l_{\text{piezo}} = 125\ \mu\text{m}$

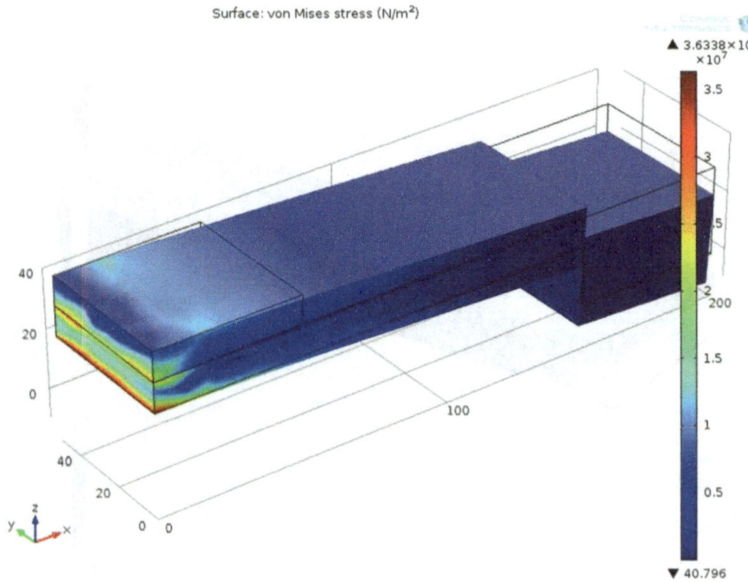

Fig. 7.45 von Mises stress when $l_{\text{piezo}} = 150\ \mu\text{m}$

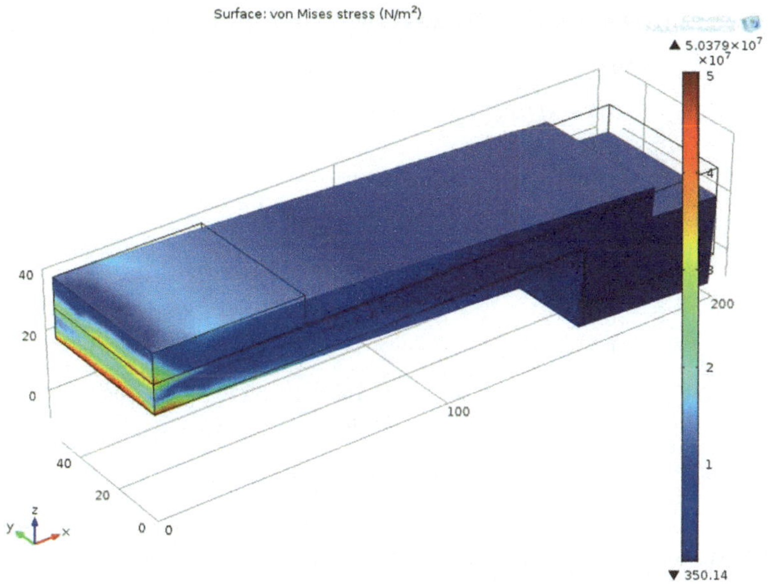

Fig. 7.46 von Mises stress when $l_{\text{piezo}} = 175\ \mu\text{m}$

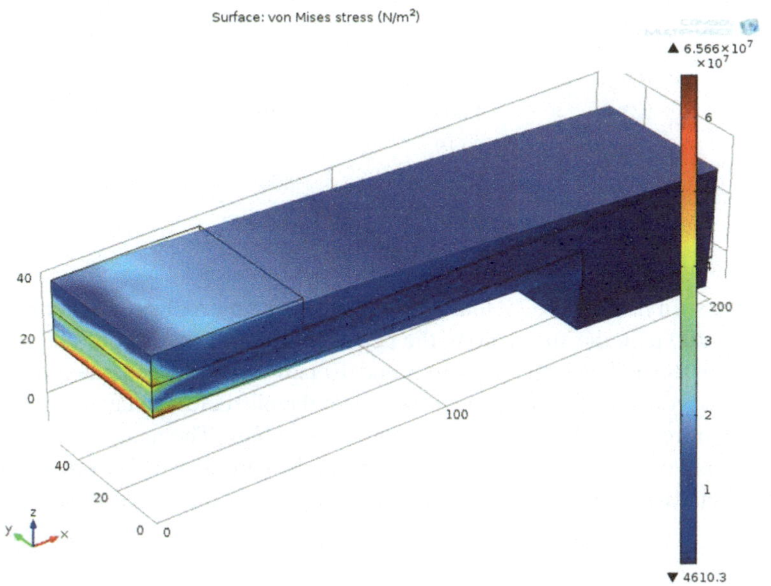

Fig. 7.47 von Mises stress when $l_{\text{piezo}} = 200\ \mu\text{m}$

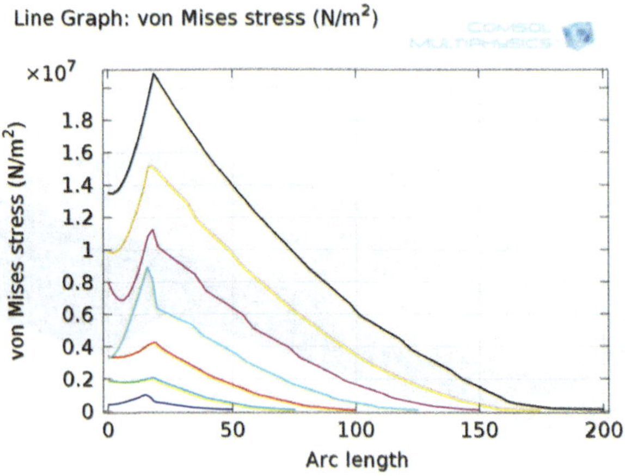

Fig. 7.48 von Mises stress versus piezoelectric layer length

Table 7.3 Maximum stress versus length for different thickness of piezoelectric layer

S. No.	l_{piezo} (μm)	von Mises stress (N/m^2) × 10^6		
		t_{piezo} = 2 μm	t_{piezo} = 5 μm	t_{piezo} = 10 μm
1	50	6.8	4.95	4.72
2	75	14.9	11.01	9.45
3	100	25.15	19.25	16.32
4	125	42.08	30.01	25.68
5	150	60.19	43.28	36.33
6	175	77.90	59.06	50.37
7	200	107.28	80.28	65.66

It can be seen from Table 7.3 that as the length of the piezoelectric layer increases from 50 to 200 μm, the stress along the beam increases. However, as the thickness of the piezoelectric layer increases to 5 and 10 μm the stress generated decreases. The stress is reduced because as the thickness of the piezoelectric layer increases the displacement reduces which results in low stress values. The analysis is important because the maximum stress on the beam should be carefully investigated specially near fixed end which must be below the fracture point of the material (silicon).

7.3.3 Electric Potential

Electric potential is generated in the piezoelectric layer due to the stress generated along the beam length. Piezoelectric layer which operates in d31 mode results in generation of charge which is collected through bottom and top electrodes. In this investigation, length (l_{piezo}) and thickness (t_{piezo}) of the piezoelectric layer are varied and electric potential is obtained. Parametric study is done by varying the length of the piezoelectric layer from 50 to 200 μm having an increment of 50 μm. Three values of the piezoelectric layer thickness, i.e., 2, 5 and 10 μm, are selected. The length of the silicon beam (L_{canti}) is 200 μm, and beam thickness (T_{canti}) is selected as 10 μm. Width of the cantilever beam L_w is 50 μm, and a seismic mass is placed at the bottom side having length 50 μm, width 50 μm and depth 30 μm.

The FEM results for the electric potential are obtained for the piezoelectric layer thickness (t_{piezo}) as 2 μm is shown below (Figs. 7.49, 7.50, 7.51, 7.52, 7.53, 7.54, 7.55 and 7.56).

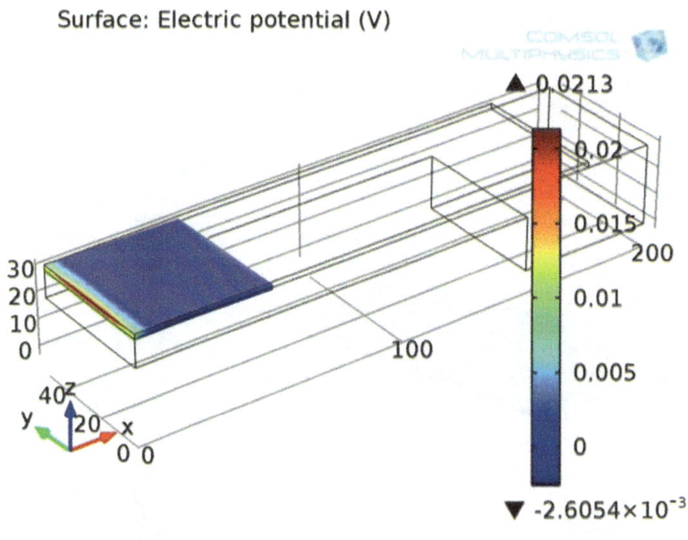

Fig. 7.49 Electric potential when $l_{piezo} = 50$ μm

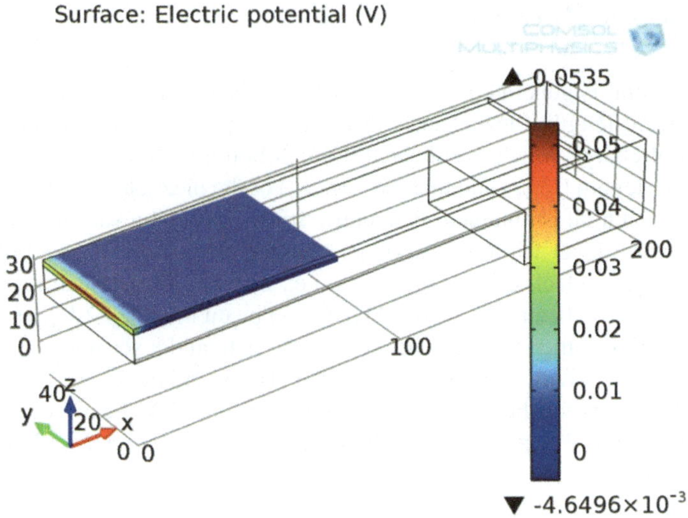

Fig. 7.50 Electric potential when $l_{\text{piezo}} = 75\ \mu\text{m}$

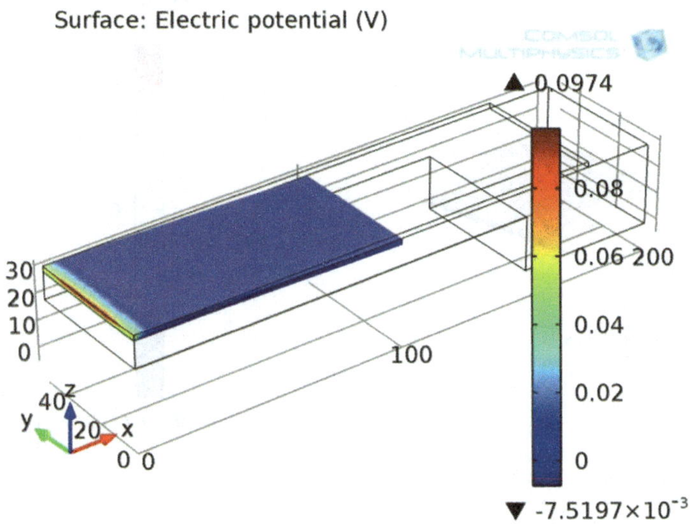

Fig. 7.51 Electric potential when $l_{\text{piezo}} = 100\ \mu\text{m}$

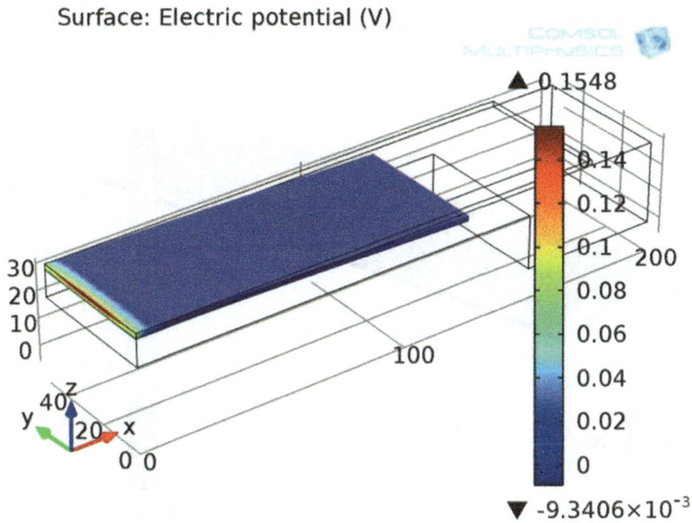

Fig. 7.52 Electric potential when $l_{piezo} = 125\ \mu m$

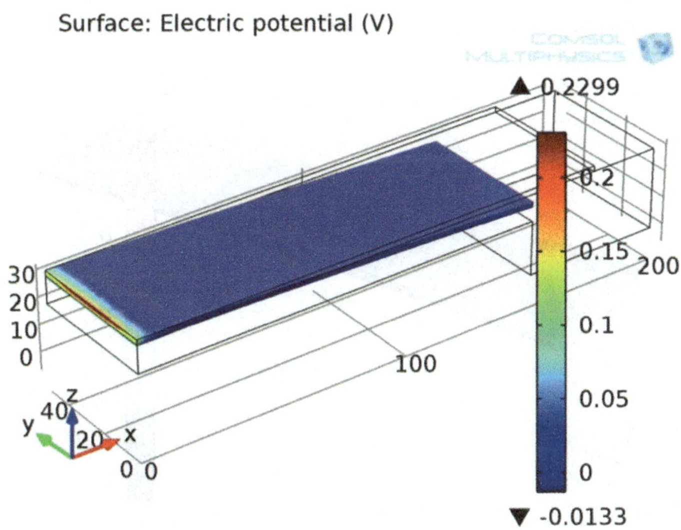

Fig. 7.53 Electric potential when $l_{piezo} = 150\ \mu m$

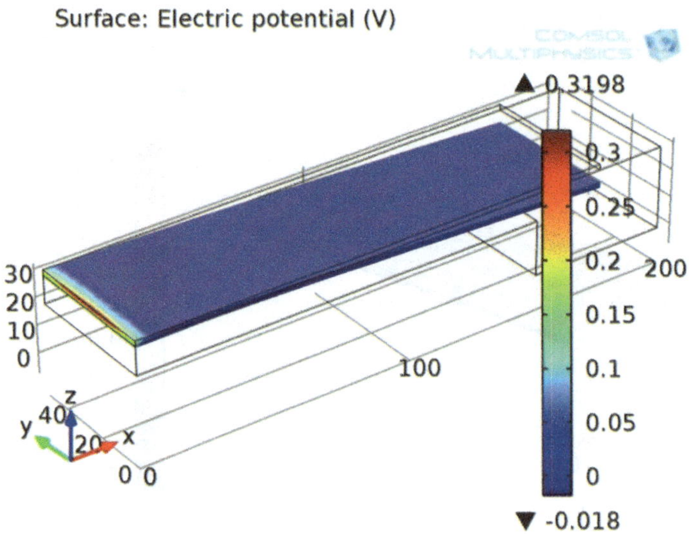

Fig. 7.54 Electric potential when $l_{piezo} = 175\ \mu m$

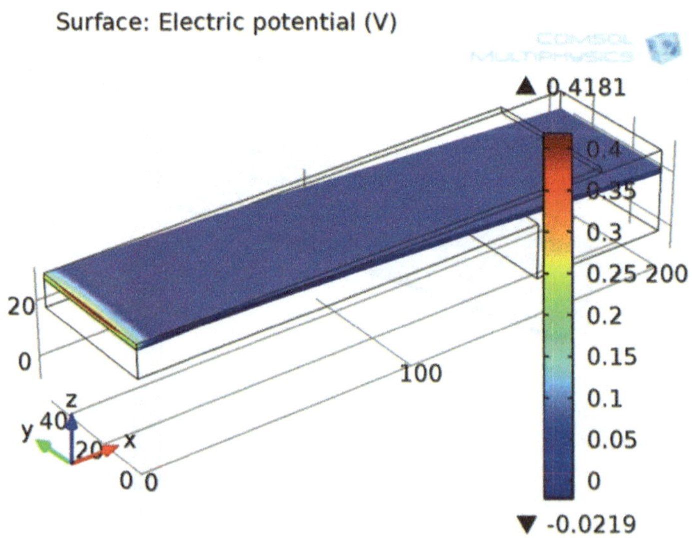

Fig. 7.55 Electric potential when $l_{piezo} = 200\ \mu m$

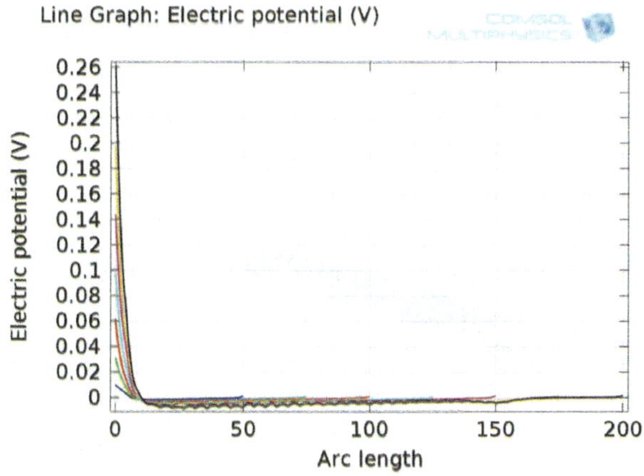

Fig. 7.56 Electric potential versus piezoelectric layer length

It can be seen from the above analysis that the electric potential is maximum at the fixed end where stress is maximum. The electric potential gradually decreases while moving from fixed end toward free end.

The FEM results for the electric potential are obtained for the piezoelectric layer thickness (t_{piezo}) as 5 μm is shown below (Figs. 7.57, 7.58, 7.59, 7.60, 7.61, 7.62, 7.63 and 7.64).

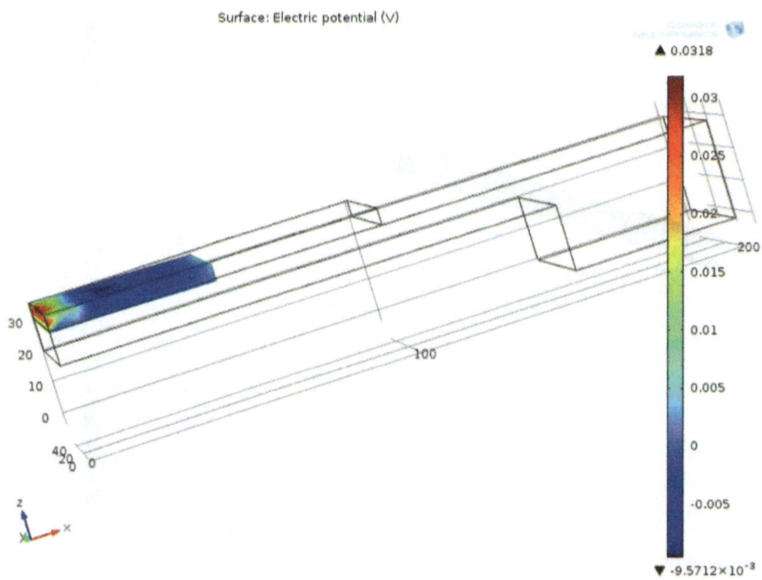

Fig. 7.57 Electric potential when $l_{piezo} = 50\ \mu m$

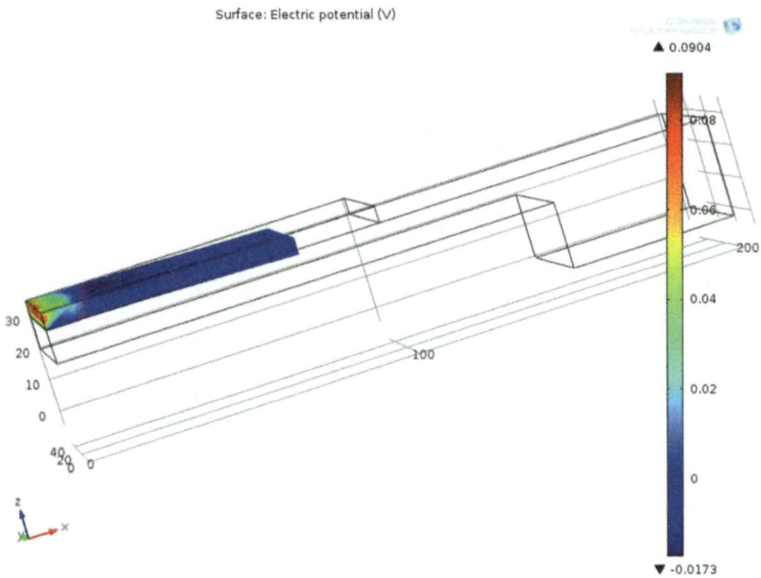

Fig. 7.58 Electric potential when $l_{\text{piezo}} = 75\ \mu\text{m}$

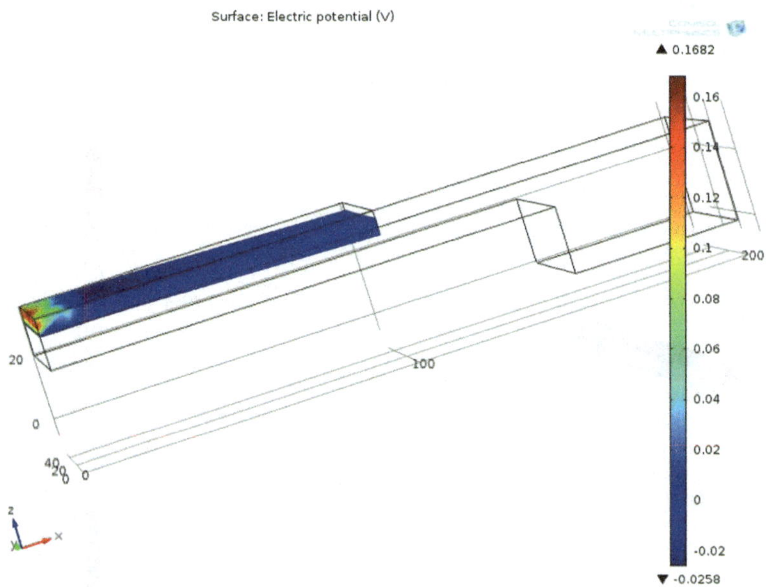

Fig. 7.59 Electric potential when $l_{\text{piezo}} = 100\ \mu\text{m}$

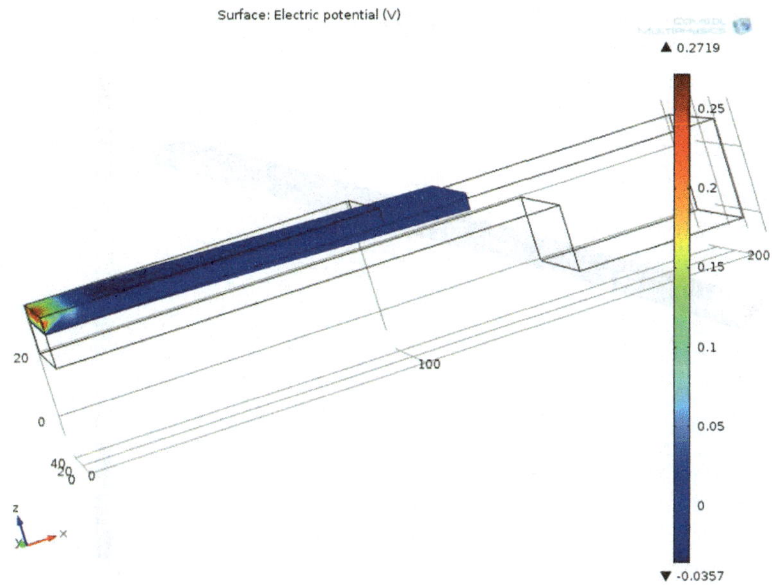

Fig. 7.60 Electric potential when $l_{\text{piezo}} = 125\ \mu\text{m}$

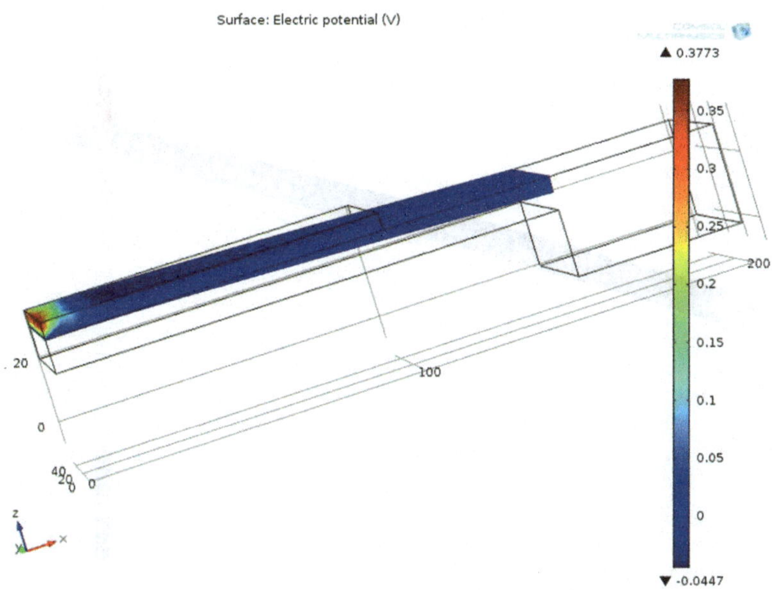

Fig. 7.61 Electric potential when $l_{\text{piezo}} = 150\ \mu\text{m}$

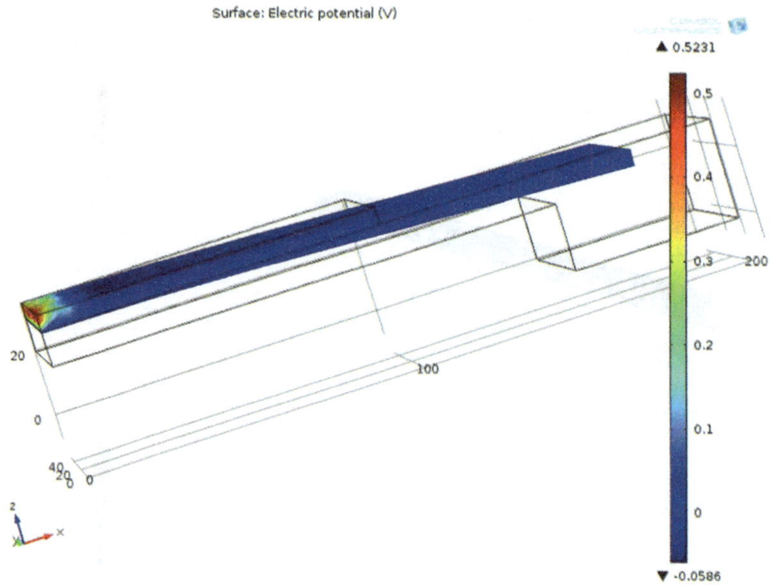

Fig. 7.62 Electric potential when $l_{\text{piezo}} = 175\ \mu\text{m}$

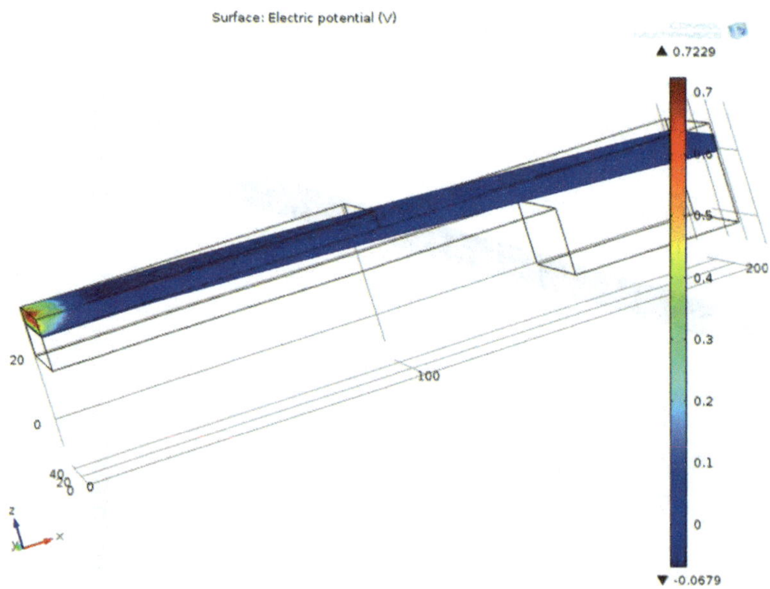

Fig. 7.63 Electric potential when $l_{\text{piezo}} = 200\ \mu\text{m}$

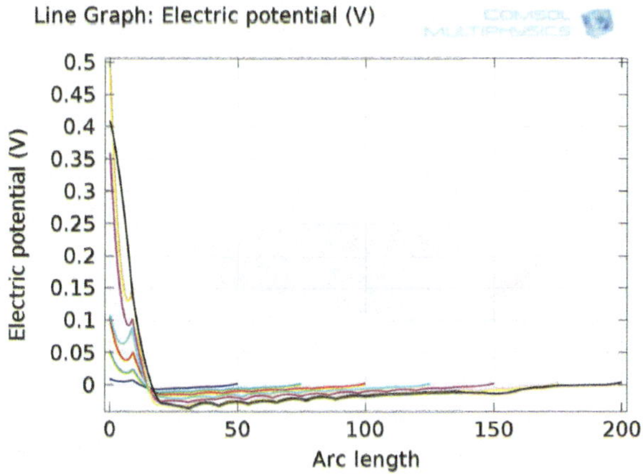

Fig. 7.64 Electric potential versus piezoelectric layer length

The FEM results for the electric potential are obtained for the piezoelectric layer thickness (t_{piezo}) as 10 μm is shown below (Figs. 7.65, 7.66, 7.67, 7.68, 7.69, 7.70, 7.71 and 7.72).

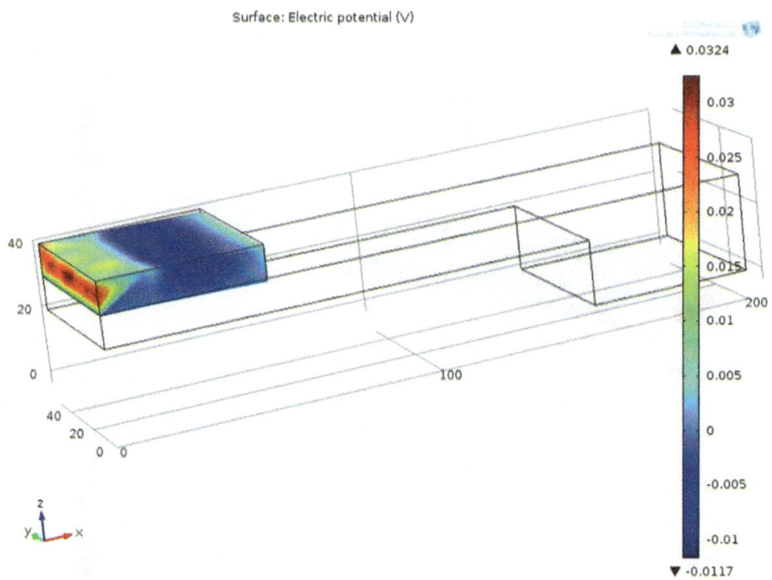

Fig. 7.65 Electric potential when $l_{\text{piezo}} = 50$ μm

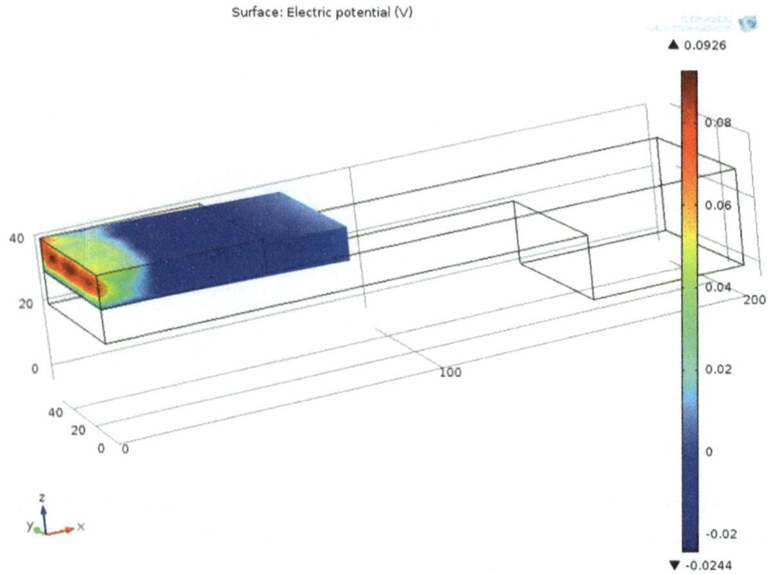

Fig. 7.66 Electric potential when $l_{piezo} = 75\ \mu m$

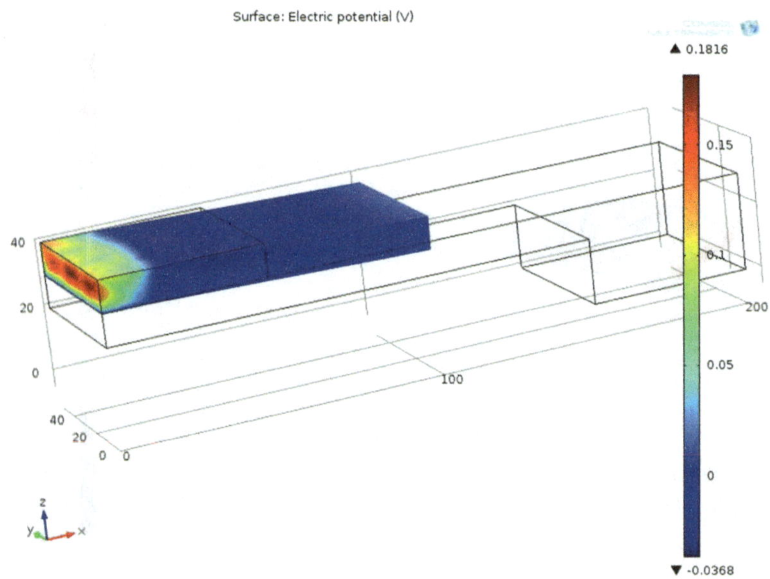

Fig. 7.67 Electric potential when $l_{piezo} = 100\ \mu m$

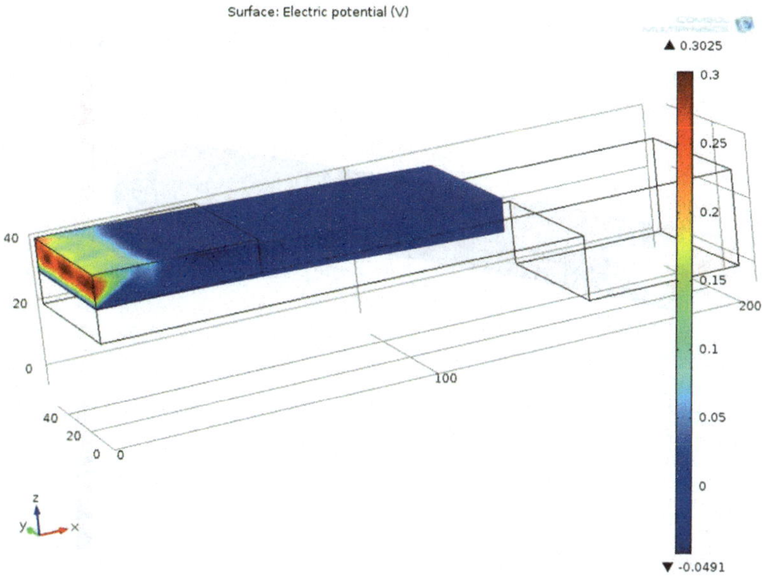

Fig. 7.68 Electric potential when $l_{\text{piezo}} = 125\ \mu\text{m}$

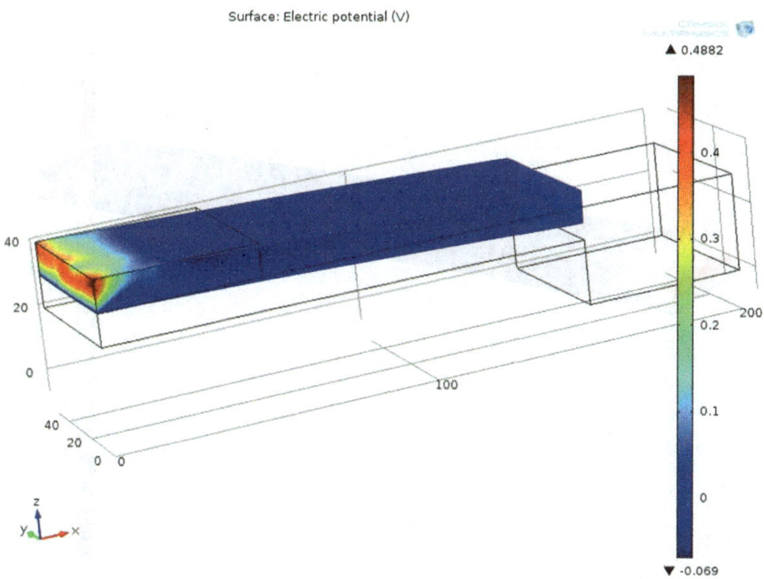

Fig. 7.69 Electric potential when $l_{\text{piezo}} = 150\ \mu\text{m}$

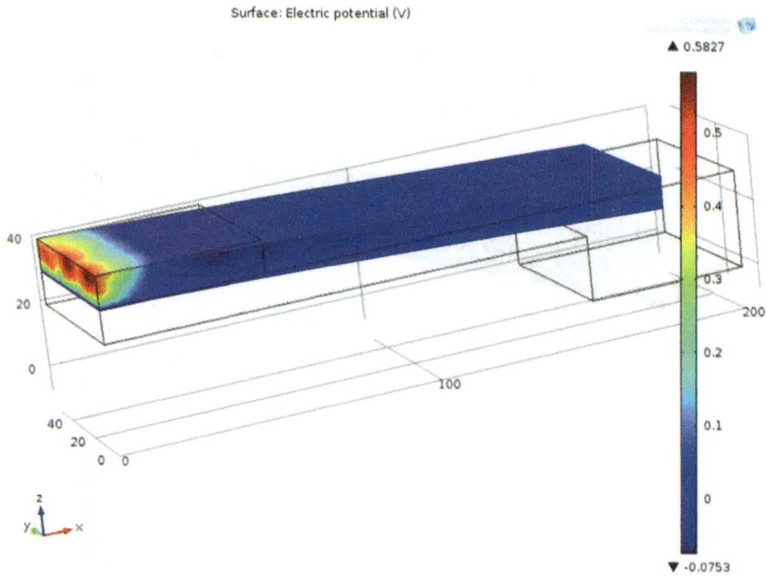

Fig. 7.70 Electric potential when $l_{\text{piezo}} = 175\ \mu\text{m}$

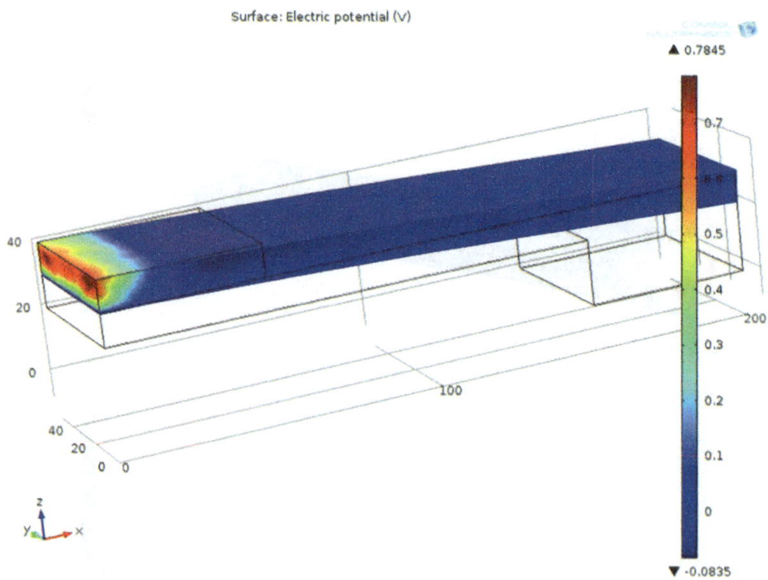

Fig. 7.71 Electric potential when $l_{\text{piezo}} = 200\ \mu\text{m}$

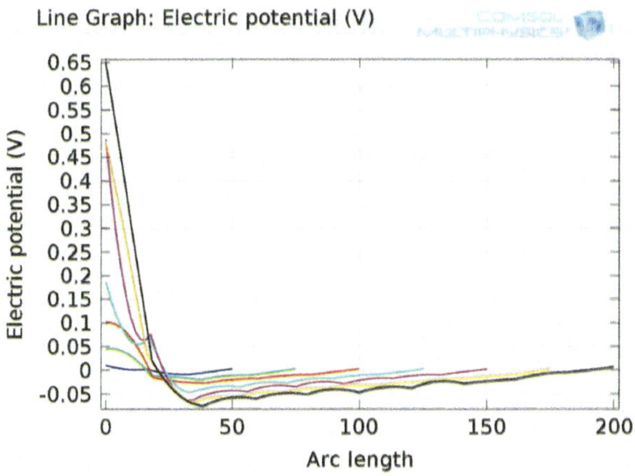

Fig. 7.72 Electric potential versus piezoelectric layer length

Table 7.4 Potential versus length at different thickness of piezoelectric layer

S. No.	l_{piezo} (μm)	Electrical potential (V)		
		$t_{\text{piezo}} = 2$ μm	$t_{\text{piezo}} = 5$ μm	$t_{\text{piezo}} = 10$ μm
1	50	0.021	0.031	0.032
2	75	0.053	0.090	0.092
3	100	0.097	0.168	0.181
4	125	0.154	0.271	0.302
5	150	0.229	0.377	0.488
6	175	0.319	0.523	0.582
7	200	0.418	0.722	0.784

It can be seen from Table 7.4 that as the length of the piezoelectric layer increases from 50 to 200 μm, the output electric potential increases from 0.021 to 0.418 V. As the thickness of the piezoelectric layer increases, the generated electric potential also increases. However, there is not much difference in the electric potential at 5 and 10 μm thickness this is because after a thickness value the piezoelectric properties gets saturated. Therefore, the thickness of the piezoelectric layer must be carefully chosen for maximum electric potential generation which is desired for energy harvesting applications.

7.4 Effect of Length and Thickness of Beam on Cantilever Type P-VEH

Thickness and length of the beam are important parameters in the device design. The thickness of the beam has an effect on the device resonance frequency. Length of the beam also affects the device resonance frequency and the size of the device.

7.4.1 Displacement

In this investigation, length of the cantilever beam (L_{canti}) is varied from 100 to 500 μm having an increment of 100 μm. Thickness (T_{canti}) of the cantilever takes three values, i.e., 5, 10 and 15 μm. Thickness (t_{piezo}) of the piezoelectric layer is fixed to 5 μm, and the length of the piezoelectric layer (l_{piezo}) changes with the length of the beam during FEM analysis. The FEM results for the cantilever displacement are obtained for the beam thickness (T_{canti}) as 5 μm is shown below. Width of the cantilever beam L_{w} is 50 μm, and a seismic mass is placed at the bottom side having length 50 μm, width 50 μm and depth 30 μm (Figs. 7.73, 7.74, 7.75, 7.76 and 7.77).

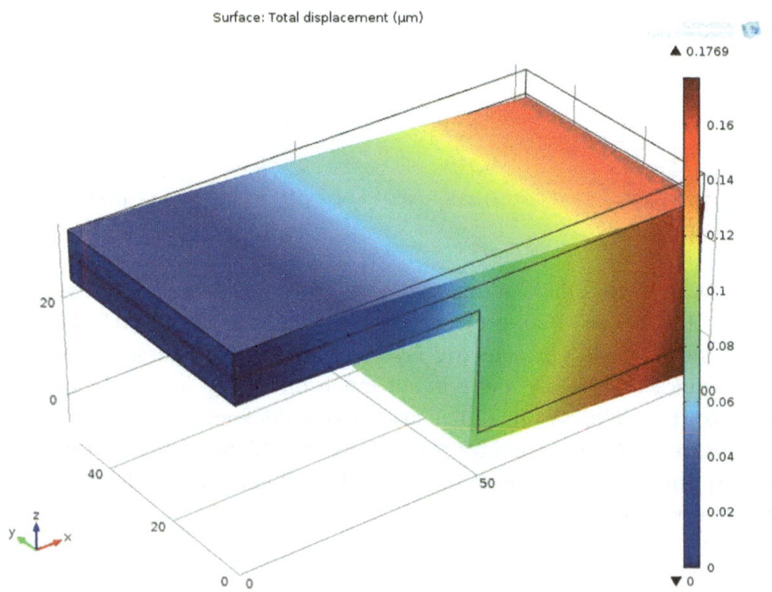

Fig. 7.73 Displacement when $L_{\text{canti}} = 100\ \mu\text{m}$

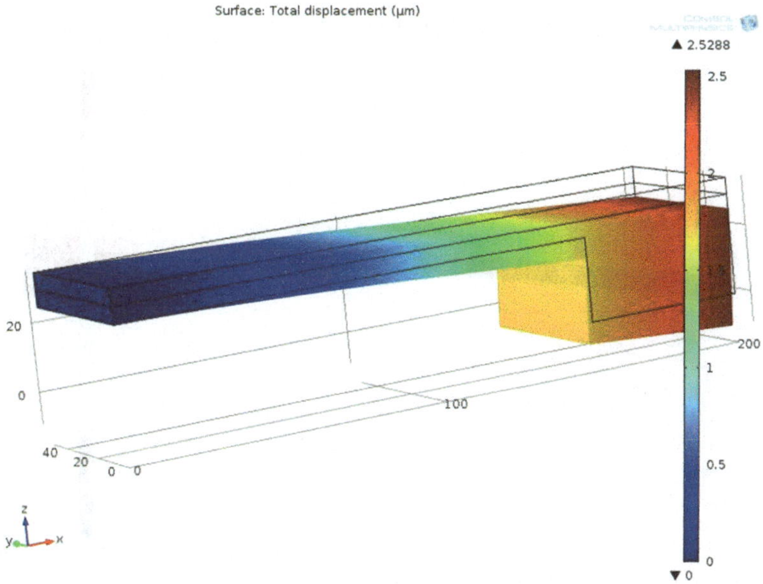

Fig. 7.74 Displacement when $L_{canti} = 200\ \mu m$

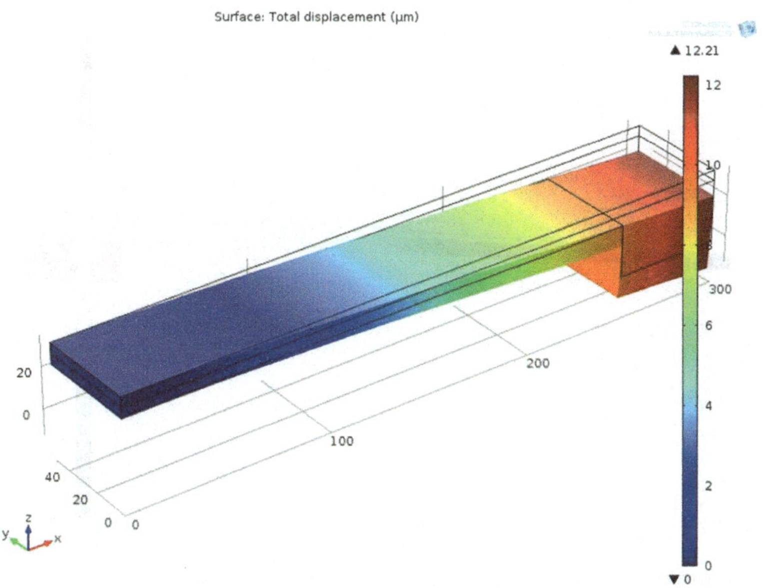

Fig. 7.75 Displacement when $L_{canti} = 300\ \mu m$

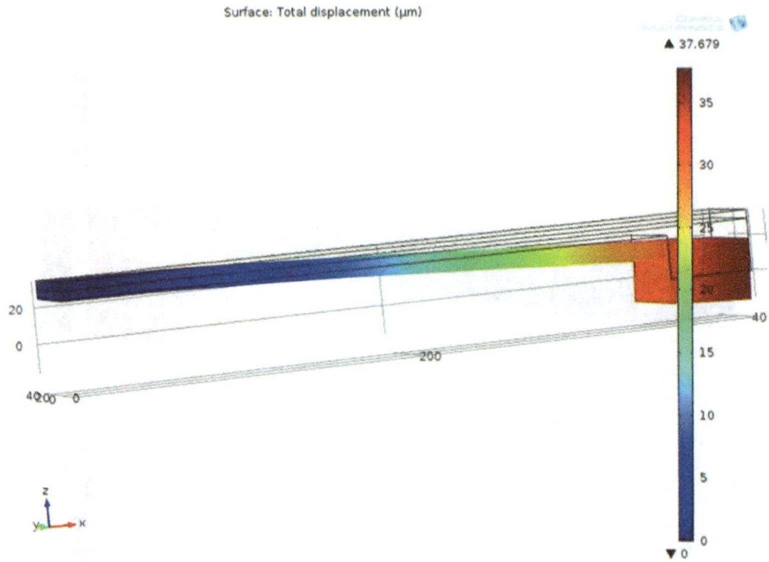

Fig. 7.76 Displacement when $L_{canti} = 400\ \mu m$

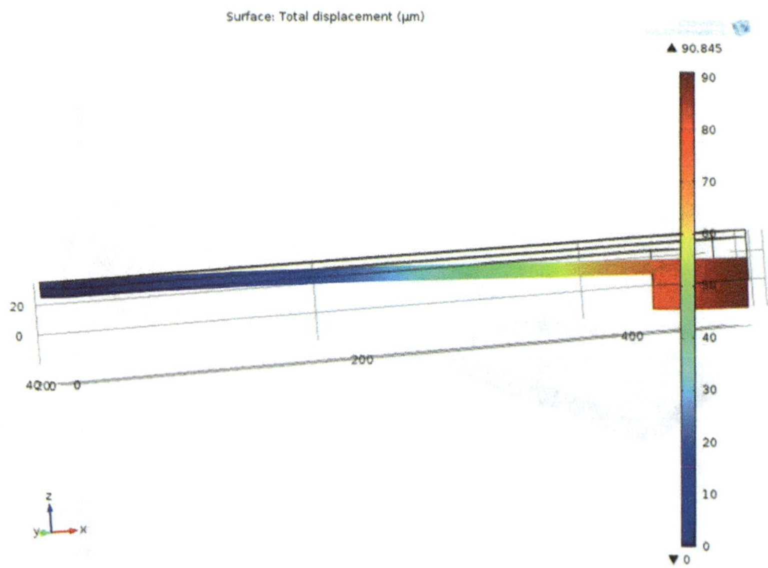

Fig. 7.77 Displacement when $L_{canti} = 500\ \mu m$

The FEM results for the cantilever displacement are obtained for the beam thickness (T_{canti}) as 10 μm is shown below (Figs. 7.78, 7.79, 7.80, 7.81 and 7.82).

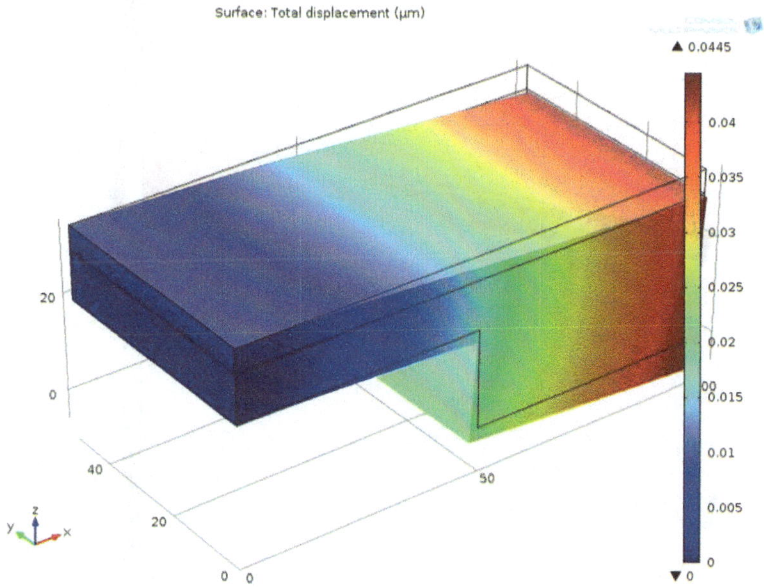

Fig. 7.78 Displacement when $L_{canti} = 100$ μm

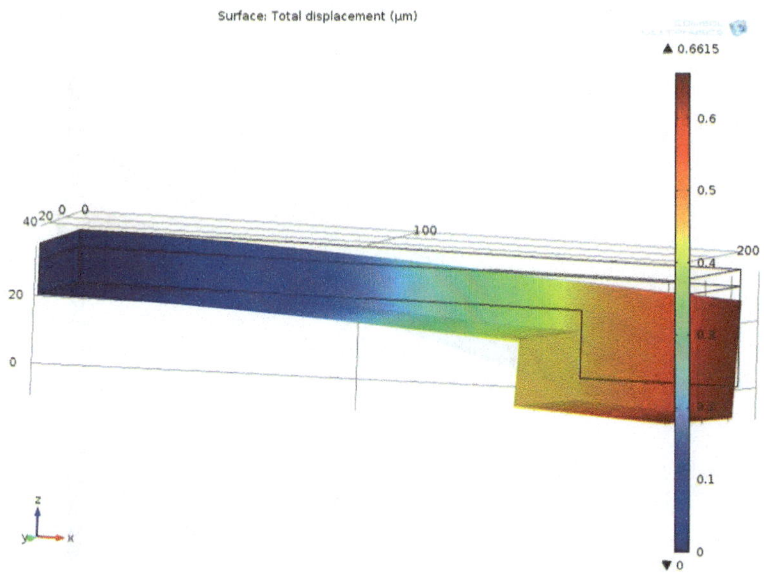

Fig. 7.79 Displacement when $L_{canti} = 200$ μm

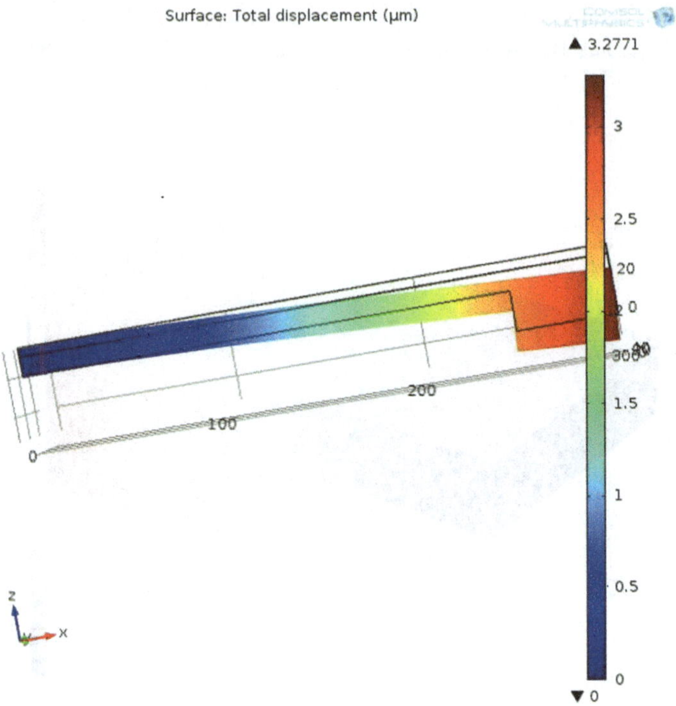

Fig. 7.80 Displacement when $L_{canti} = 300\ \mu m$

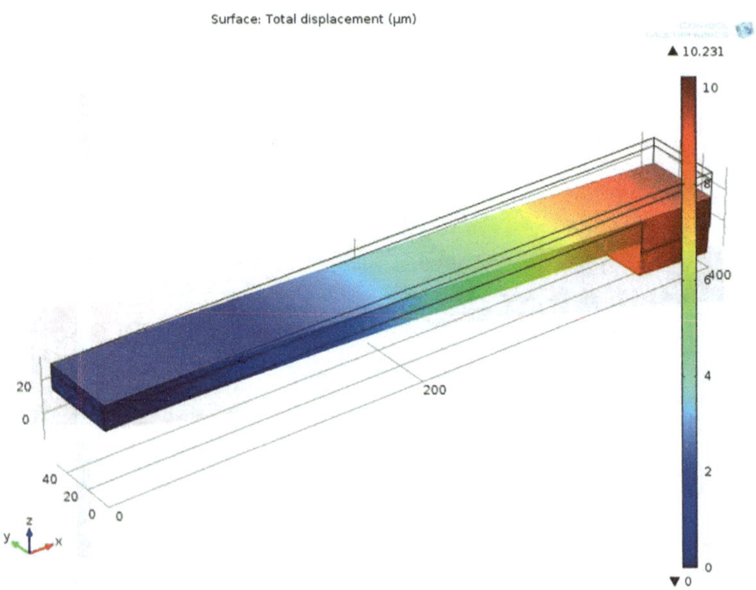

Fig. 7.81 Displacement when $L_{canti} = 400\ \mu m$

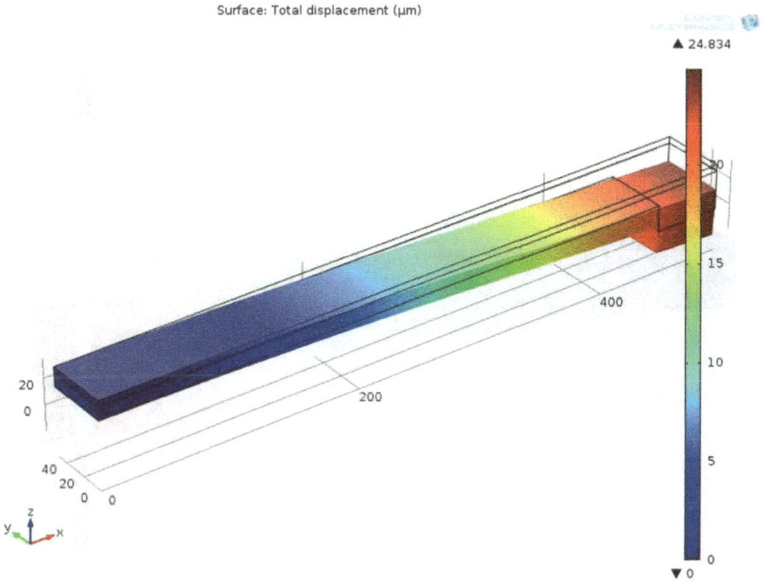

Fig. 7.82 Displacement when $L_{\text{canti}} = 500\ \mu\text{m}$

The FEM results for the cantilever displacement are obtained for the beam thickness (T_{canti}) as 15 µm is shown below (Figs. 7.83, 7.84, 7.85, 7.86, 7.87 and 7.88).

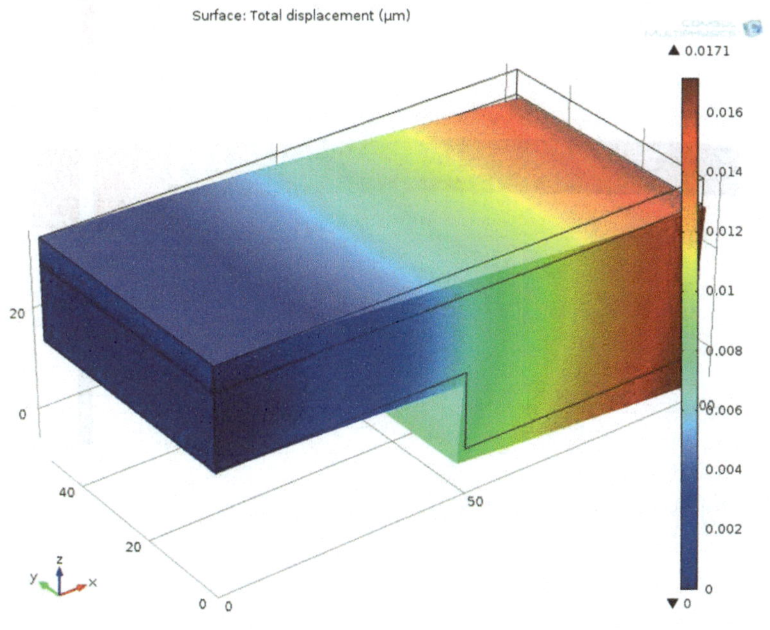

Fig. 7.83 Displacement when $L_{\text{canti}} = 100\ \mu\text{m}$

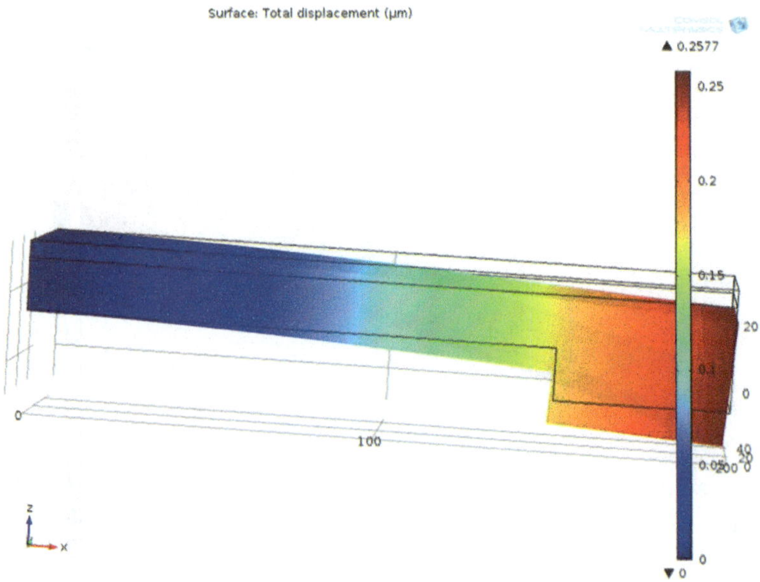

Fig. 7.84 Displacement when $L_{canti} = 200\ \mu m$

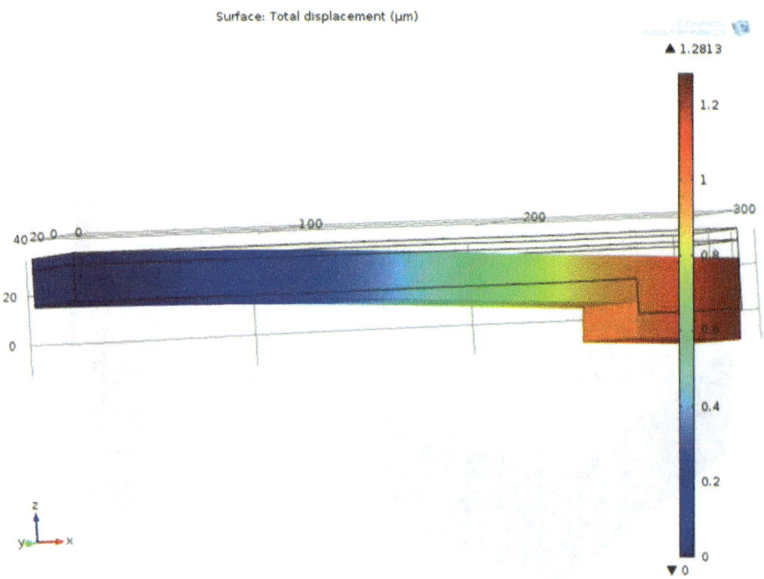

Fig. 7.85 Displacement when $L_{canti} = 300\ \mu m$

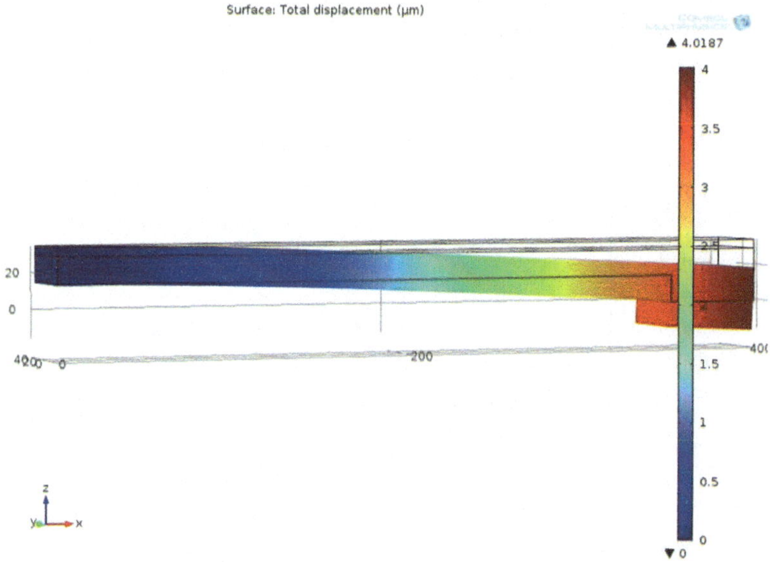

Fig. 7.86 Displacement when $L_{\text{canti}} = 400\ \mu\text{m}$

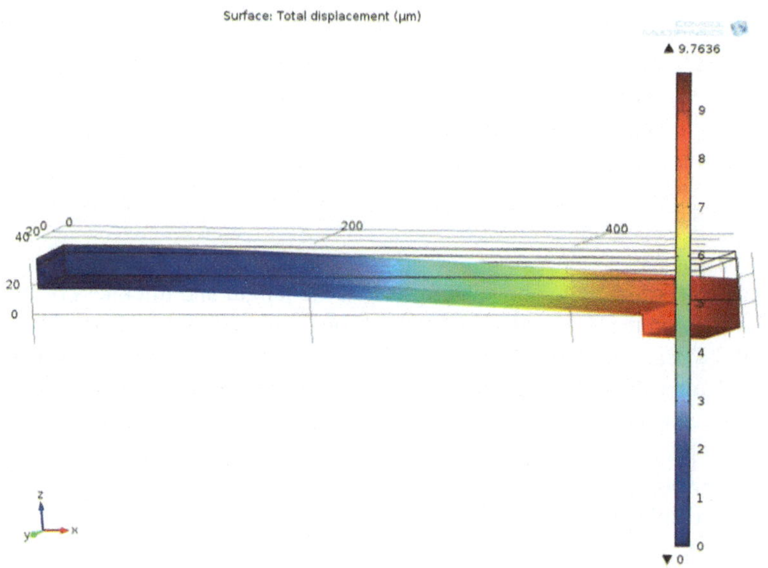

Fig. 7.87 Displacement when $L_{\text{canti}} = 500\ \mu\text{m}$

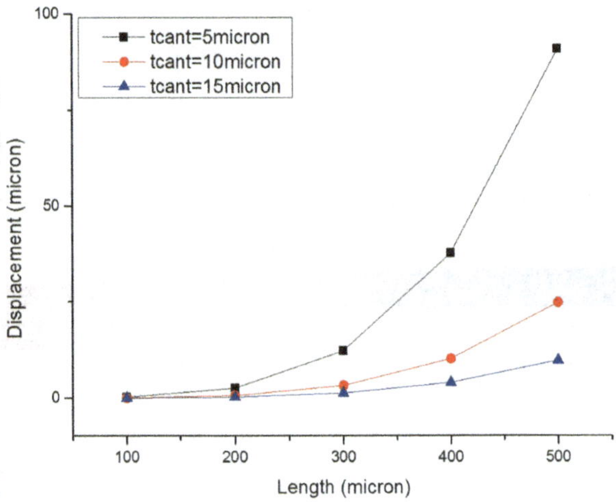

Fig. 7.88 Displacement versus cantilever beam length

Table 7.5 Displacement versus length of cantiliver at various thickness

S. No.	L_{canti} (μm)	Displacement (μm)		
		$T_{canti} = 5$ μm	$T_{canti} = 10$ μm	$T_{canti} = 15$ μm
1	100	0.17	0.04	0.01
2	200	2.52	0.66	0.25
3	300	12.21	3.27	1.28
4	400	37.67	10.23	4.01
5	500	90.84	24.83	9.76

It can be seen from the above analysis that the length and thickness of the beam have an effect on the device displacement. From Table 7.5, it is observed that as the length of the beam increases from 100 to 500 μm, the displacement of the cantilever increases from 0.17 to 90.84 μm for a beam thickness of 5 μm. It can also be seen that as the beam thickness increases to 10 and 15 μm, the displacement of the beam decreases. This is because as the thickness of the silicon beam increases it becomes stiff and the spring constant increases which increases the resonance frequency and reduces the displacement.

7.4.2 von Mises Stress

In this investigation, length of the cantilever beam (L_{canti}) is varied from 100 to 500 μm having an increment of 100 μm. Thickness (T_{canti}) of the cantilever takes three values, i.e., 5, 10 and 15 μm. Thickness (t_{piezo}) of the piezoelectric layer is fixed to 5 μm, and the length of the piezoelectric layer (l_{piezo}) changes with the length of the beam during FEM analysis. Width of the cantilever beam L_w is 50 μm, and a seismic mass is placed at the bottom side having length 50 μm, width 50 μm and depth 30 μm. The FEM results for the von Mises stress are obtained for the beam thickness (T_{canti}) as 5 μm is shown below (Figs. 7.89, 7.90, 7.91, 7.92 and 7.93).

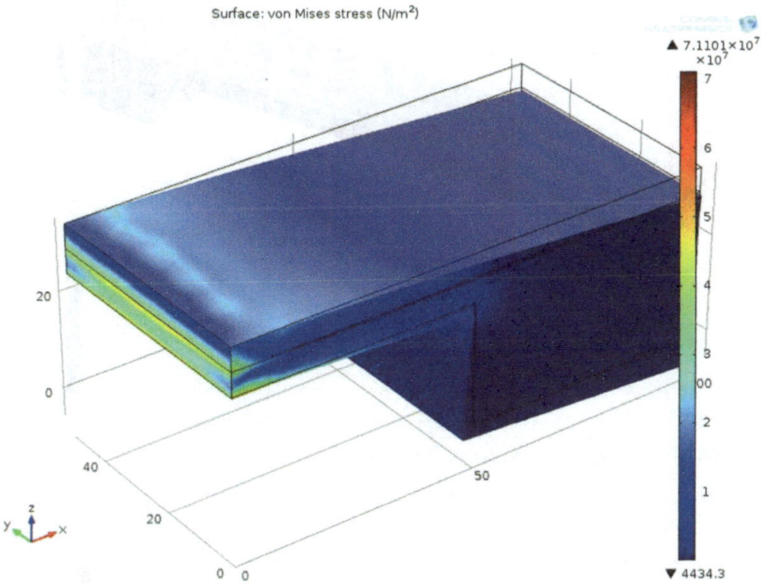

Fig. 7.89 von Mises stress when $L_{canti} = 100$ μm

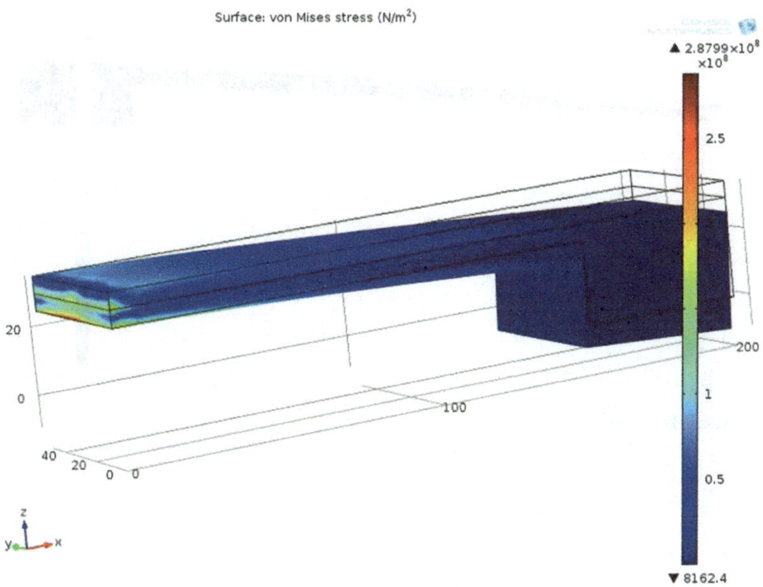

Fig. 7.90 von Mises stress when $L_{canti} = 200$ μm

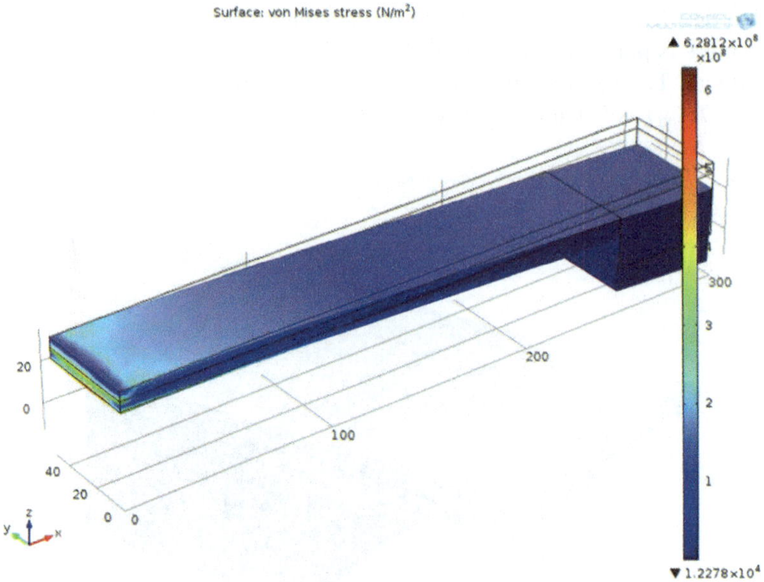

Fig. 7.91 von Mises stress when $L_{\text{canti}} = 300\ \mu\text{m}$

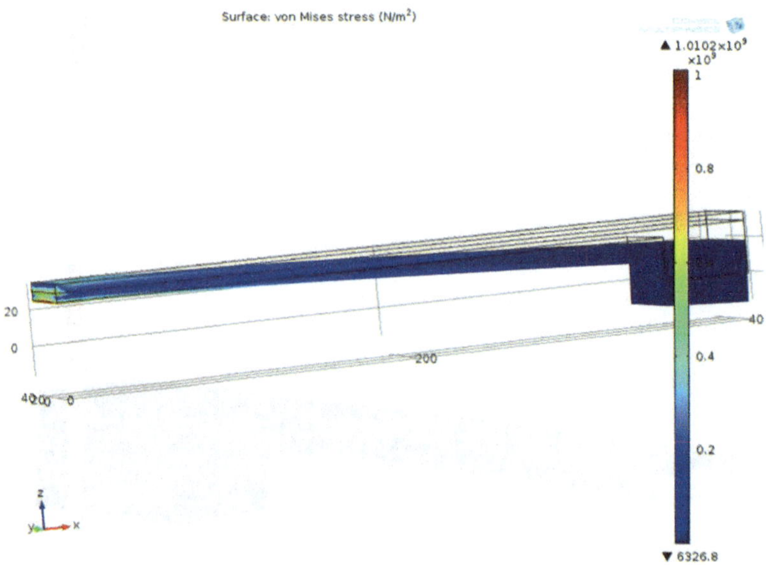

Fig. 7.92 von Mises stress when $L_{\text{canti}} = 400\ \mu\text{m}$

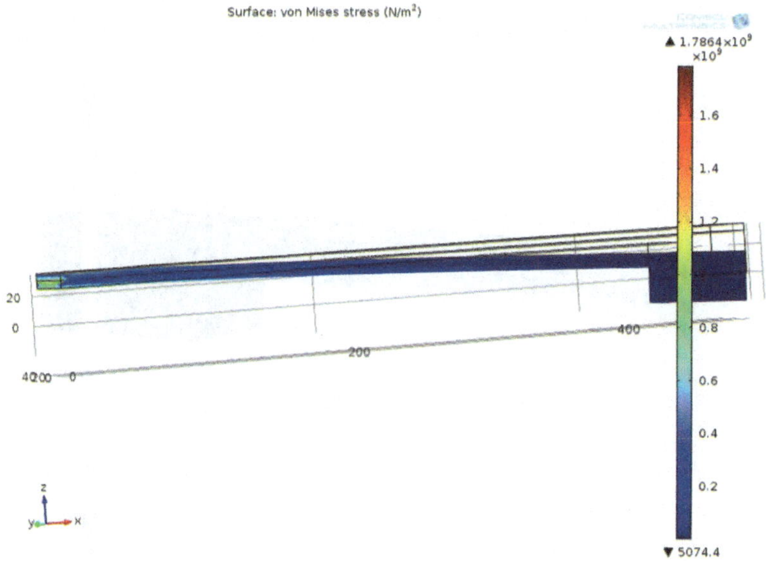

Fig. 7.93 von Mises stress when $L_{\text{canti}} = 500\ \mu\text{m}$

The FEM results for the von Mises stress are obtained for the beam thickness (T_{canti}) as 10 μm is shown below (Figs. 7.94, 7.95, 7.96, 7.97 and 7.98).

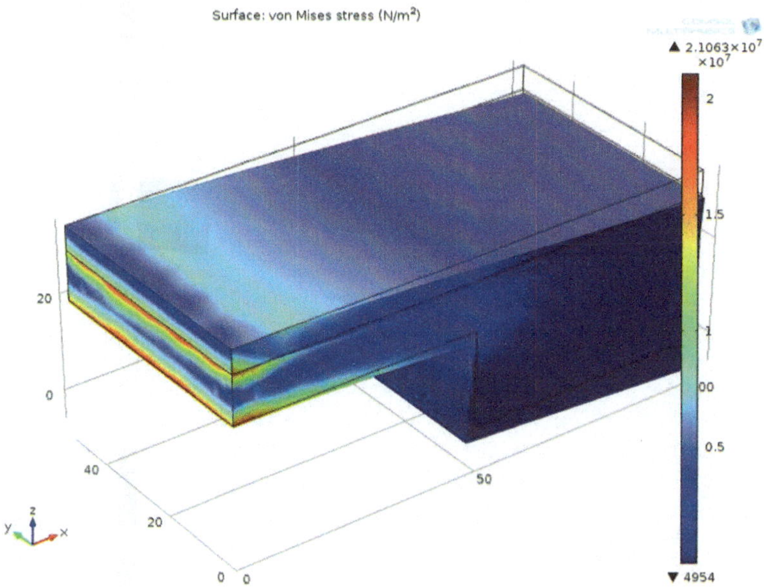

Fig. 7.94 von Mises stress when $L_{\text{canti}} = 100\ \mu\text{m}$

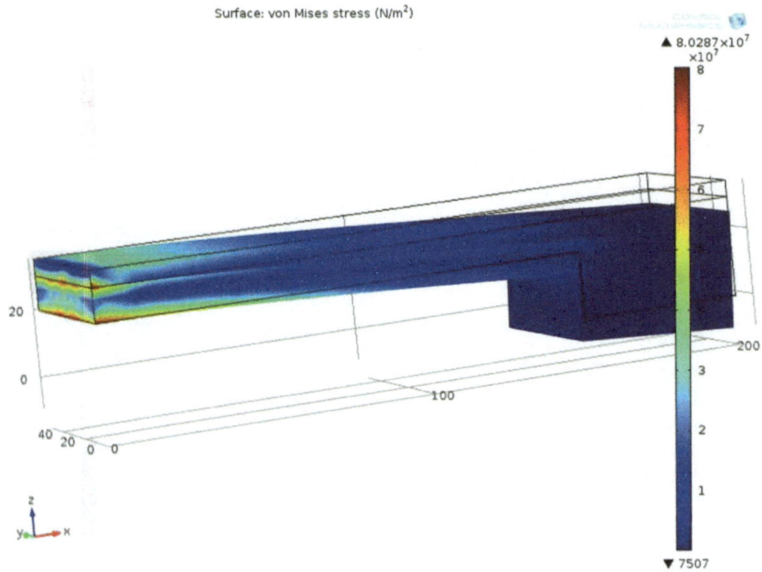

Fig. 7.95 von Mises stress when $L_{\mathrm{canti}} = 200\ \mu\mathrm{m}$

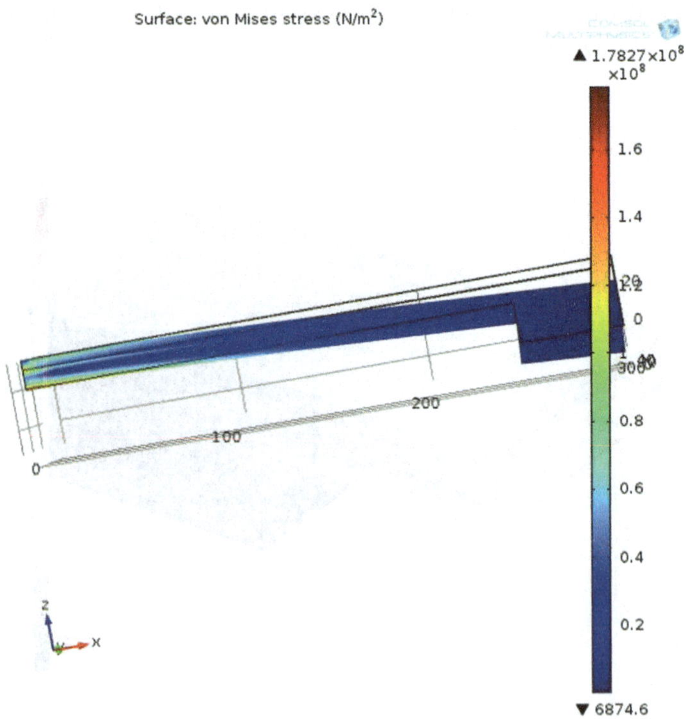

Fig. 7.96 von Mises stress when $L_{\mathrm{canti}} = 300\ \mu\mathrm{m}$

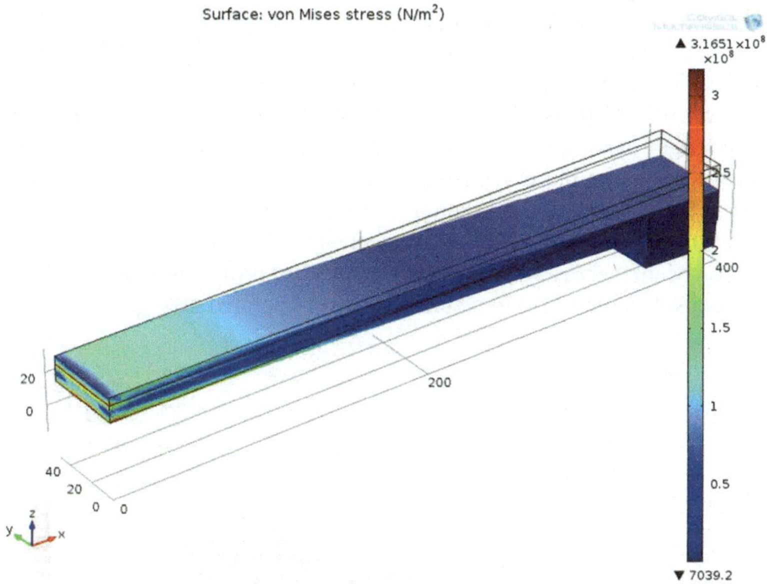

Fig. 7.97 von Mises stress when $L_{canti} = 400$ μm

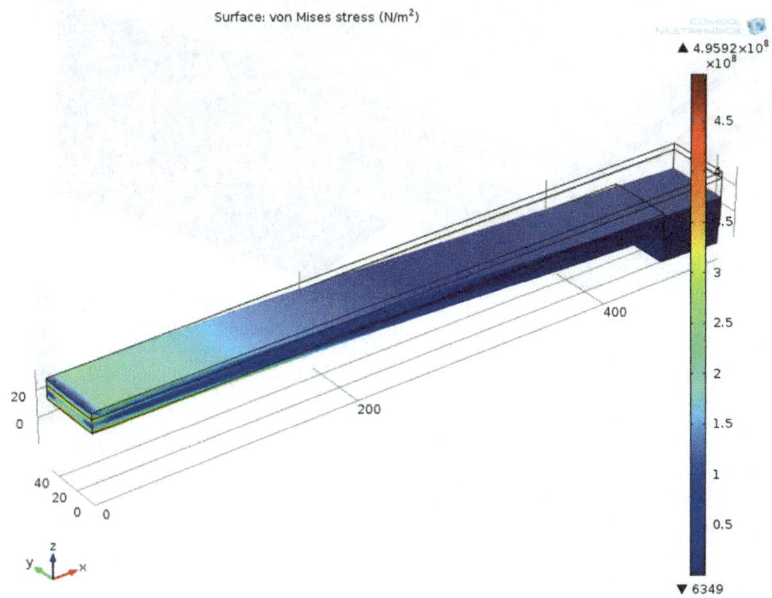

Fig. 7.98 von Mises stress when $L_{canti} = 500$ μm

The FEM results for the von Mises stress are obtained for the beam thickness (T_{canti}) as 15 μm is shown below (Figs. 7.99, 7.100, 7.101, 7.102, 7.103 and 7.104).

It can be seen from the above analysis that the length and thickness of the beam have an effect on the stress developed on the device. From Table 7.6, it is observed that as the length of the beam increases from 100 to 500 μm, the stress developed on the cantilever beam increases from 71.1×10^6 to 580×10^6 N/m^2 for a beam thickness of 5 μm. It can also be seen that as the beam thickness increases to 10 and 15 μm, the stress generated on the beam decreases. This is because as the thickness of the silicon beam increases it becomes stiff and the spring constant increases which reduces the displacement and in result the stress generated decreases.

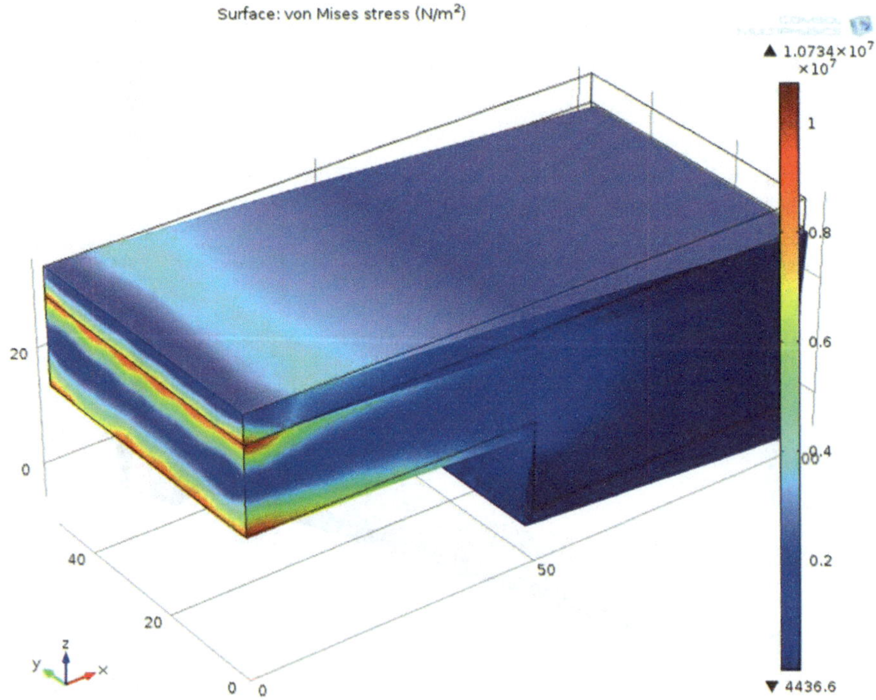

Fig. 7.99 von Mises stress when $L_{\text{canti}} = 100$ μm

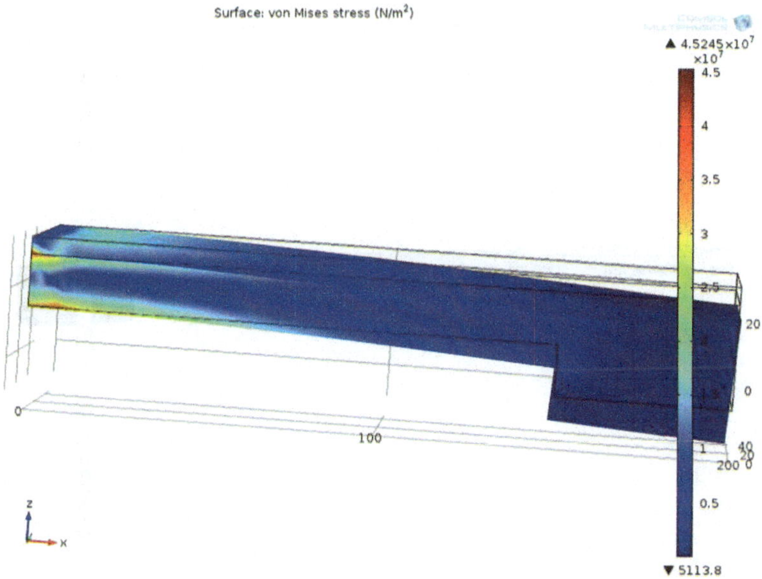

Fig. 7.100 von Mises stress when $L_{\text{canti}} = 200\ \mu\text{m}$

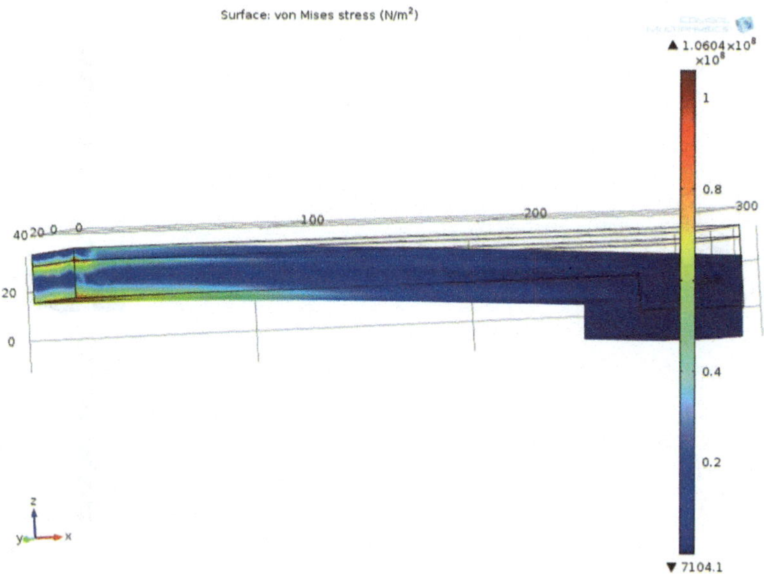

Fig. 7.101 von Mises stress when $L_{\text{canti}} = 300\ \mu\text{m}$

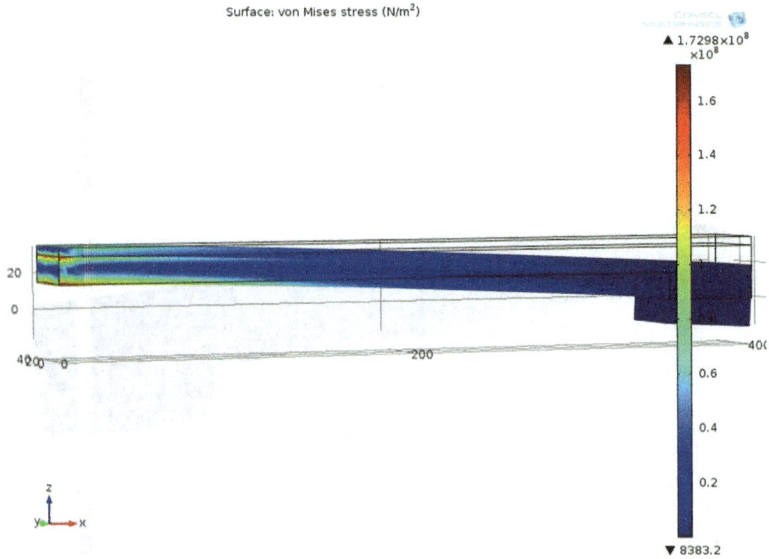

Fig. 7.102 von Mises stress when $L_{\text{canti}} = 400\ \mu\text{m}$

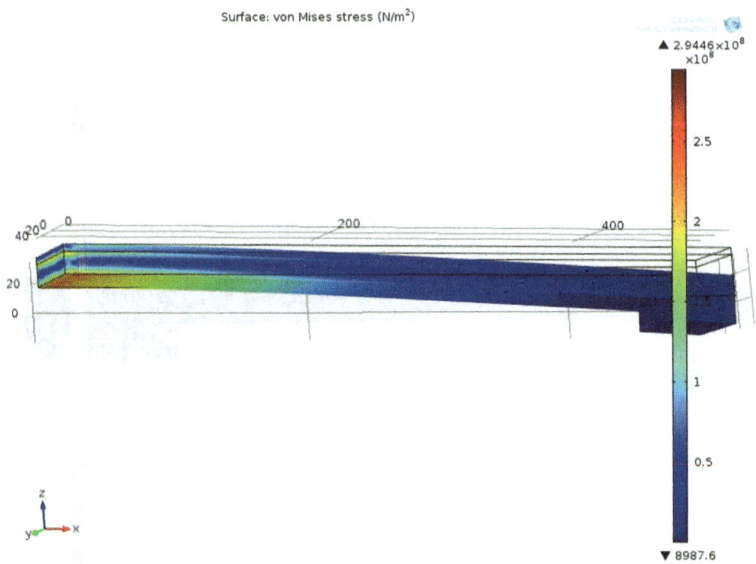

Fig. 7.103 von Mises stress when $L_{\text{canti}} = 500\ \mu\text{m}$

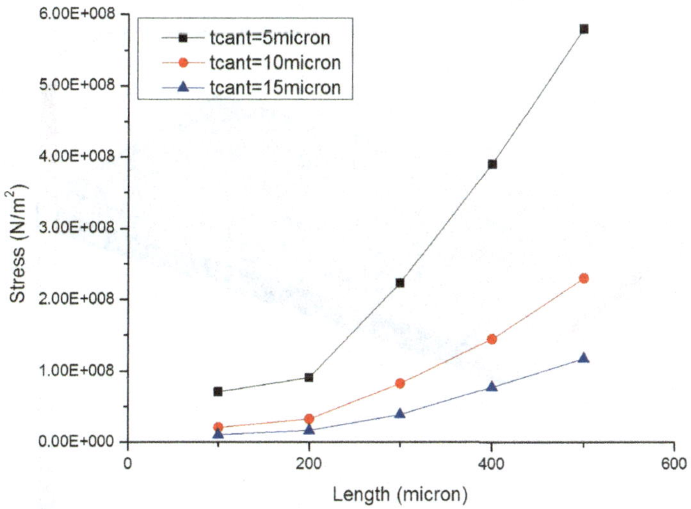

Fig. 7.104 von Mises stress versus cantilever beam length

Table 7.6 Stress versus cantilever length at different thickness

S. No.	L_{canti} (μm)	von Mises stress (N/m^2) × 10^6		
		$T_{canti} = 5$ μm	$T_{canti} = 10$ μm	$T_{canti} = 15$ μm
1	100	71.1	21.06	10.7
2	200	91.0	33	16.8
3	300	223	83	39
4	400	390	145	78
5	500	580	230	118

7.4.3 Electric Potential

In this investigation, length of the cantilever beam (L_{canti}) is varied from 100 to 500 μm having an increment of 100 μm. Thickness (T_{canti}) of the cantilever takes three values, i.e., 5, 10 and 15 μm. Thickness (t_{piezo}) of the piezoelectric layer is fixed to 5 μm, and the length of the piezoelectric layer (l_{piezo}) changes with the length of the beam during FEM analysis. Width of the cantilever beam L_w is 50 μm, and a seismic mass is placed at the bottom side having length 50 μm, width 50 μm and depth 30 μm.

The FEM results for the electric potential are obtained for the beam thickness (T_{canti}) as 5 μm is shown below (Figs. 7.105, 7.106, 7.107, 7.108 and 7.109).

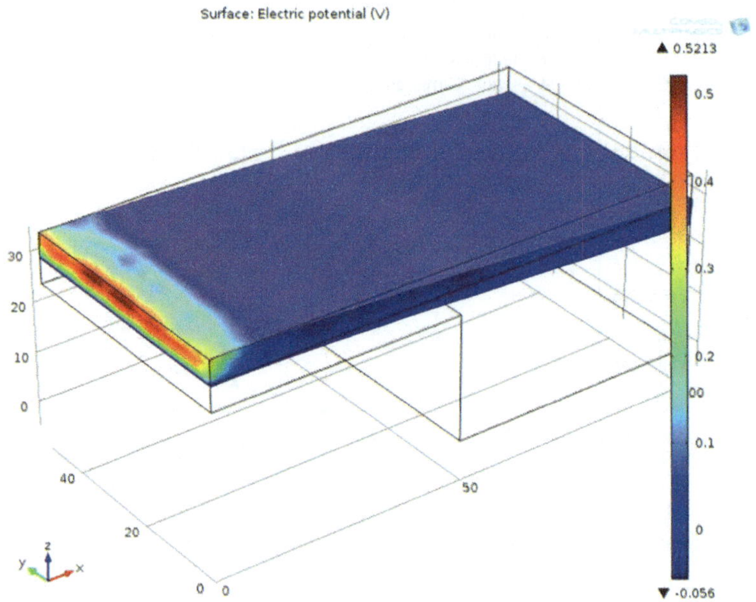

Fig. 7.105 Electric potential when $L_{canti} = 100\ \mu m$

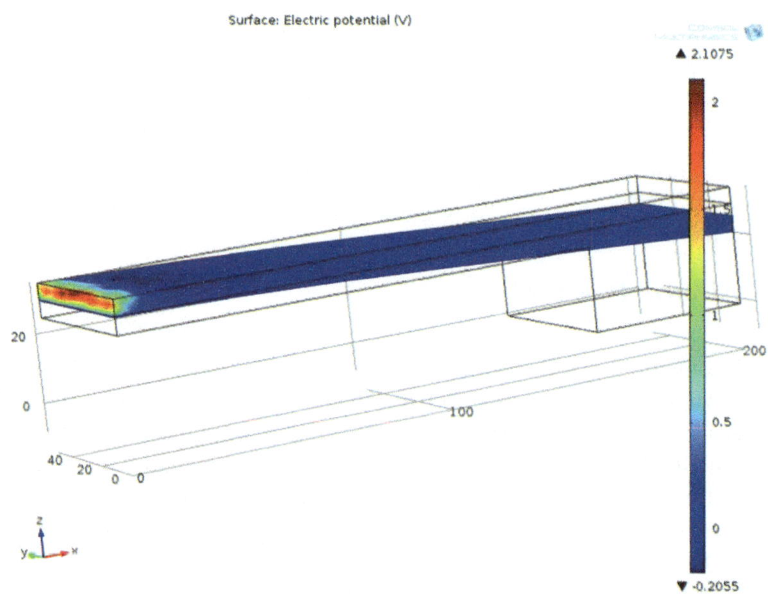

Fig. 7.106 Electric potential when $L_{canti} = 200\ \mu m$

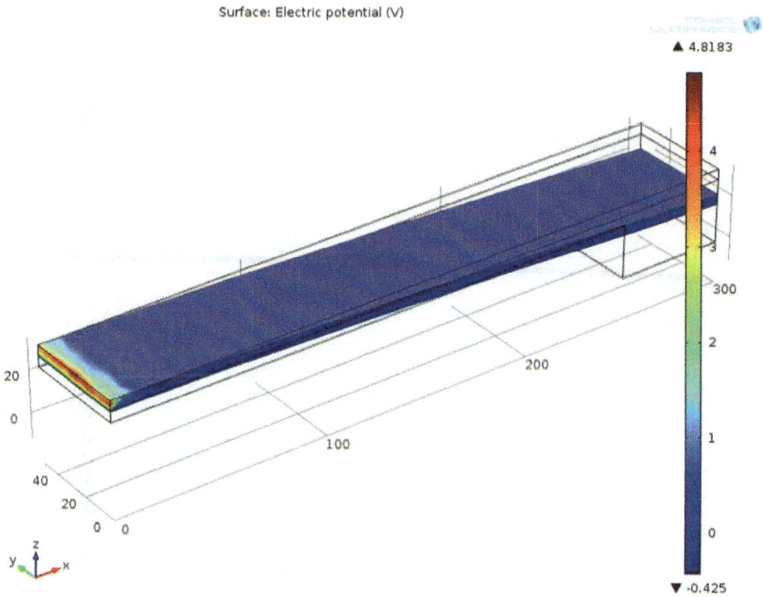

Fig. 7.107 Electric potential when $L_{\text{canti}} = 300\ \mu\text{m}$

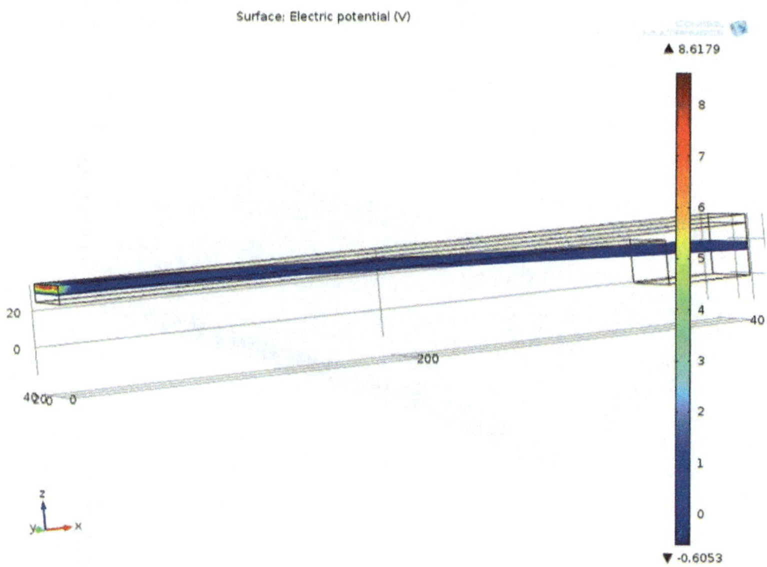

Fig. 7.108 Electric potential when $L_{\text{canti}} = 400\ \mu\text{m}$

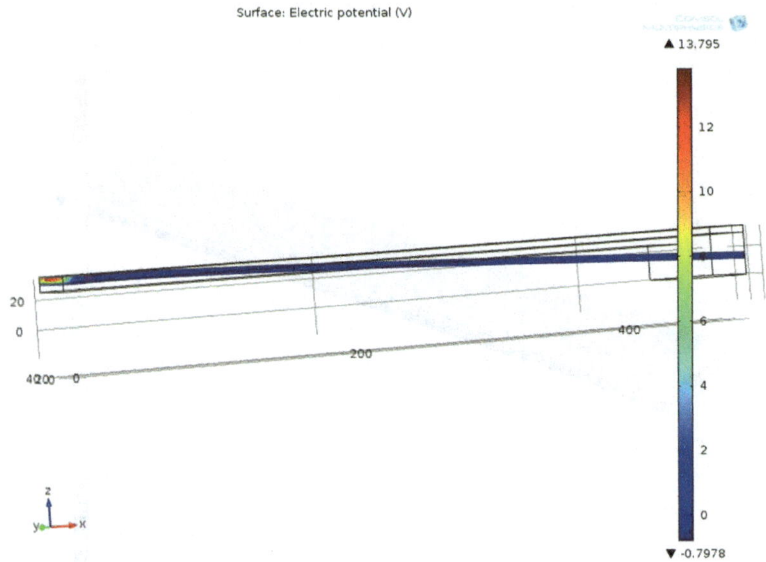

Fig. 7.109 Electric potential when $L_{\text{canti}} = 500\ \mu\text{m}$

The FEM results for the electric potential are obtained for the beam thickness (T_{canti}) as 10 μm is shown below (Figs. 7.110, 7.111, 7.112, 7.113 and 7.114).

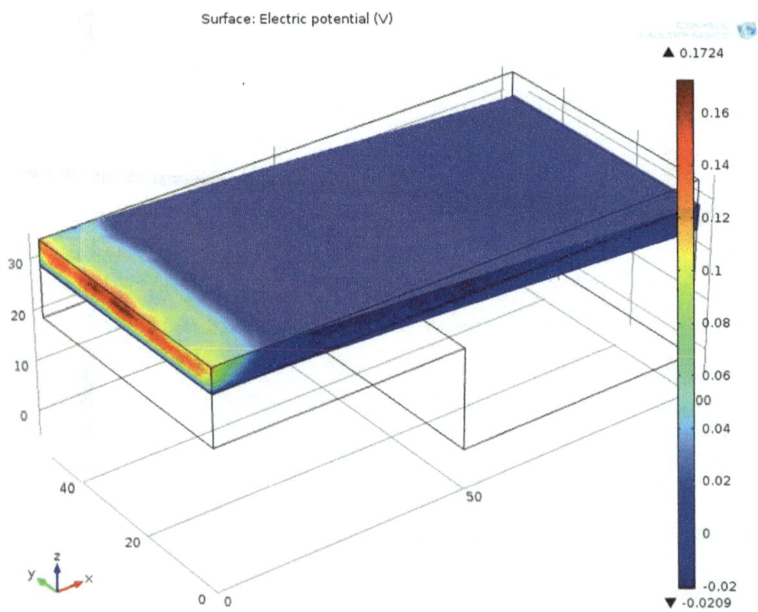

Fig. 7.110 Electric potential when $L_{\text{canti}} = 100\ \mu\text{m}$

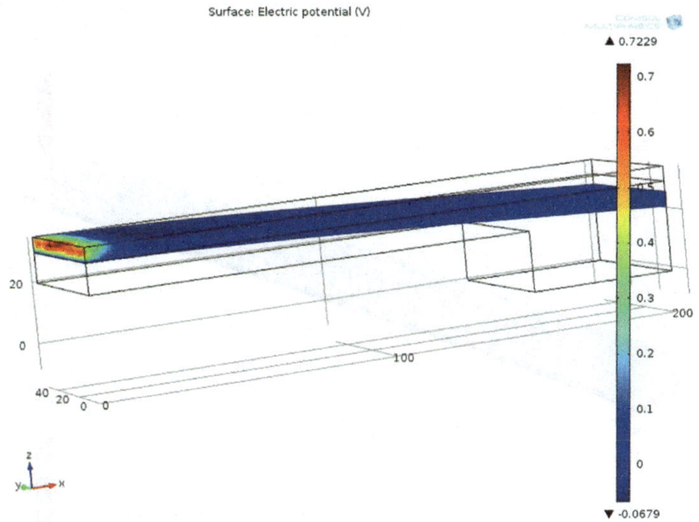

Fig. 7.111 Electric potential when $L_{\mathrm{canti}} = 200\ \mu\mathrm{m}$

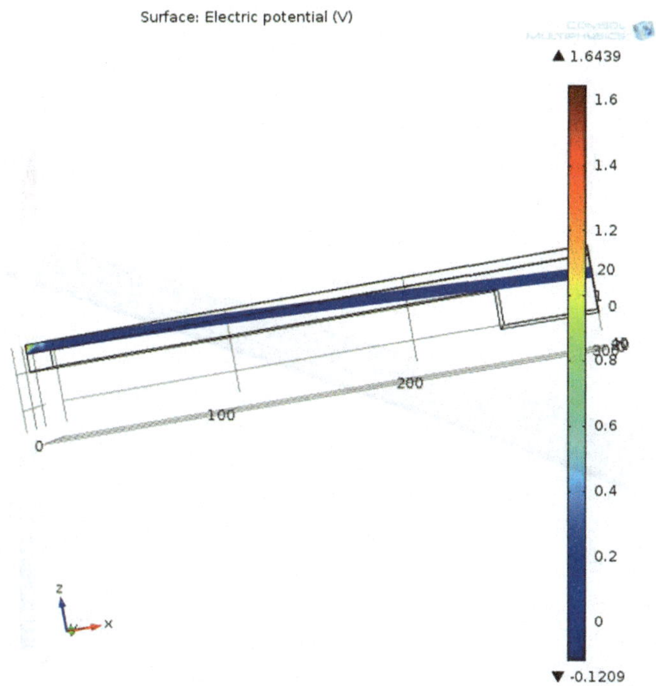

Fig. 7.112 Electric potential when $L_{\mathrm{canti}} = 300\ \mu\mathrm{m}$

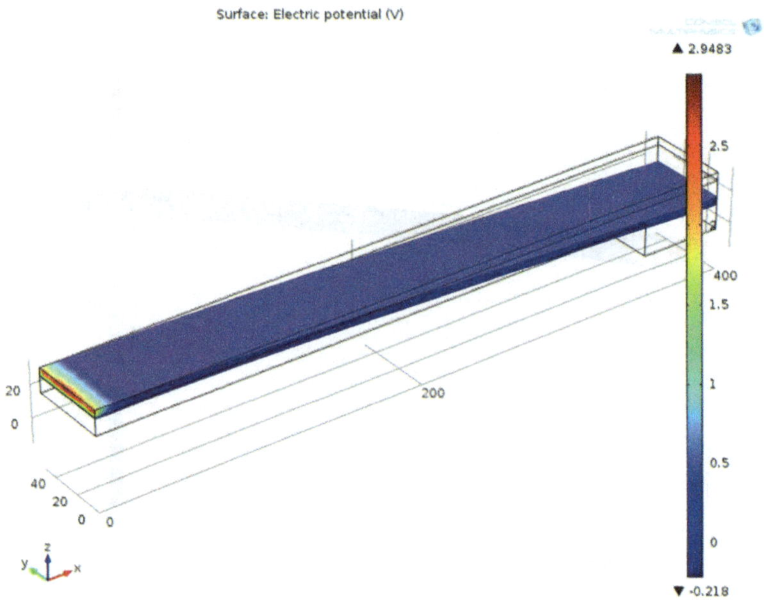

Fig. 7.113 Electric potential when $L_{canti} = 400\ \mu m$

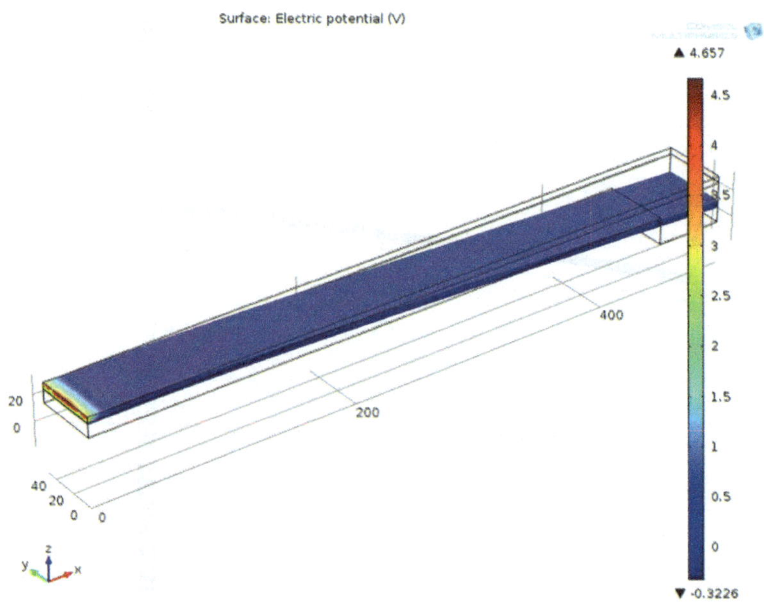

Fig. 7.114 Electric potential when $L_{canti} = 500\ \mu m$

The FEM results for the electric potential are obtained for the beam thickness (T_{canti}) as 15 μm is shown below (Figs. 7.115, 7.116, 7.117, 7.118, 7.119 and 7.120).

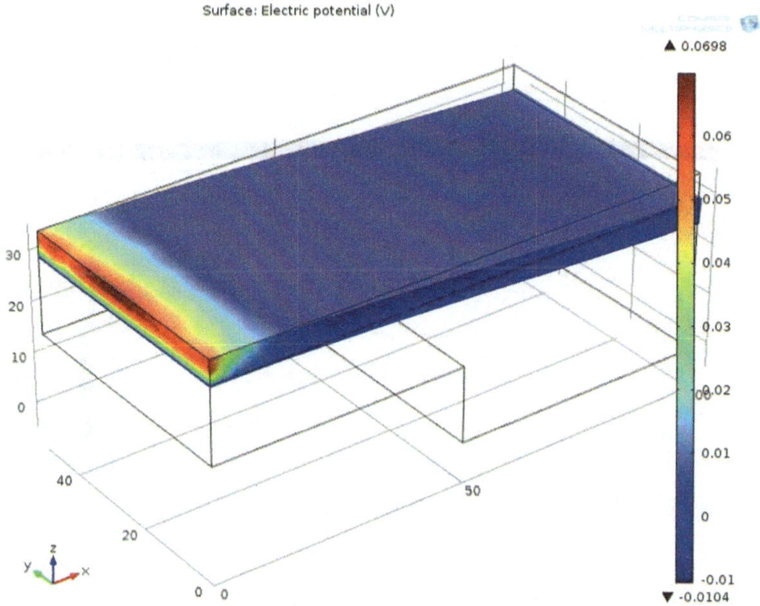

Fig. 7.115 Electric potential when $L_{canti} = 100$ μm

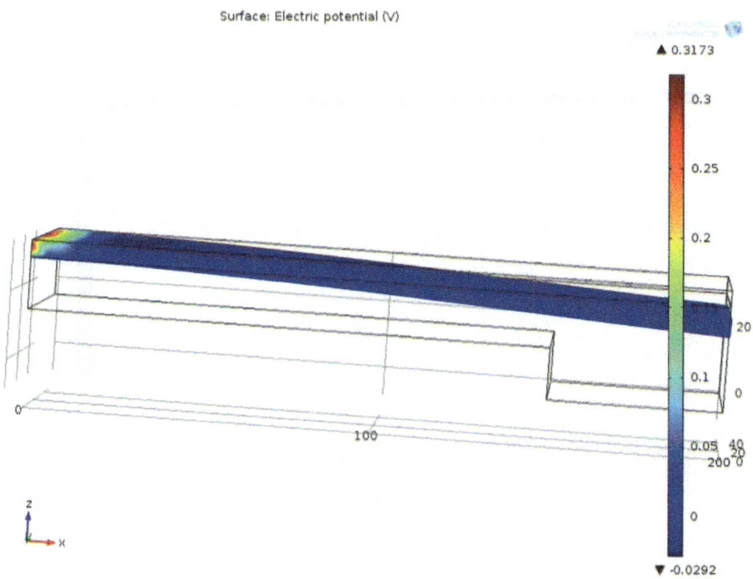

Fig. 7.116 Electric potential when $L_{canti} = 200$ μm

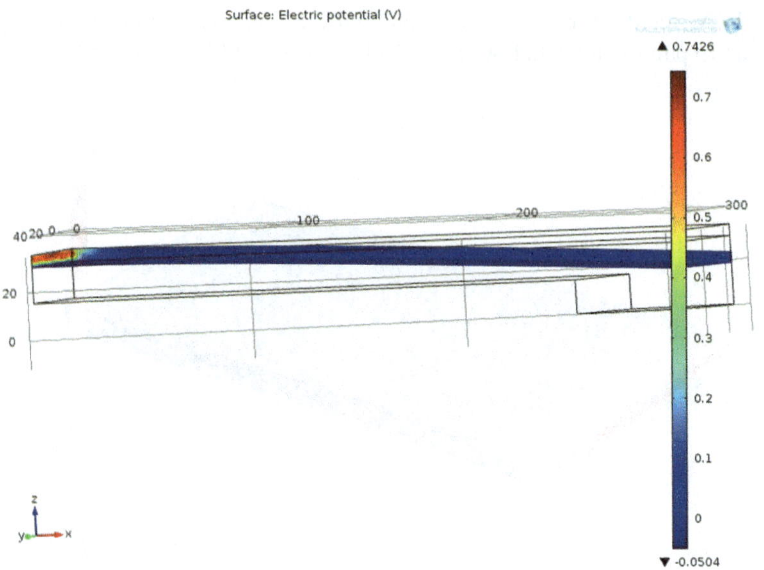

Fig. 7.117 Electric potential when $L_{canti} = 300\ \mu m$

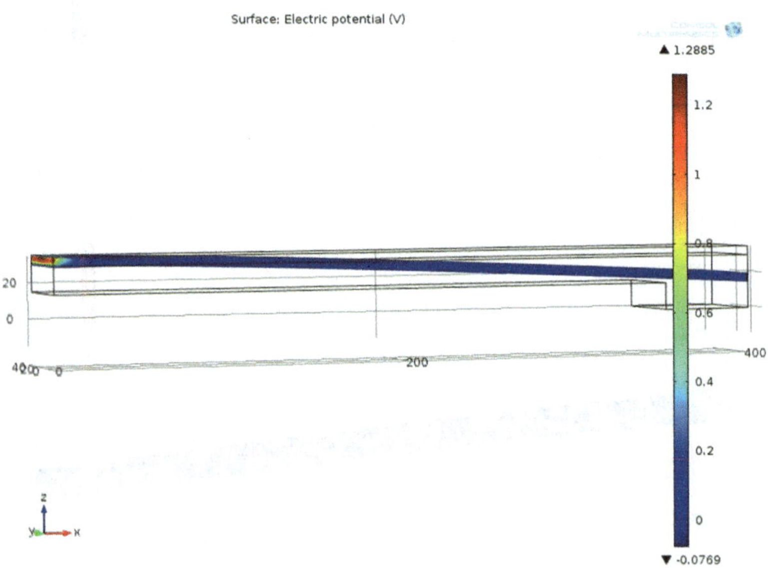

Fig. 7.118 Electric potential when $L_{canti} = 400\ \mu m$

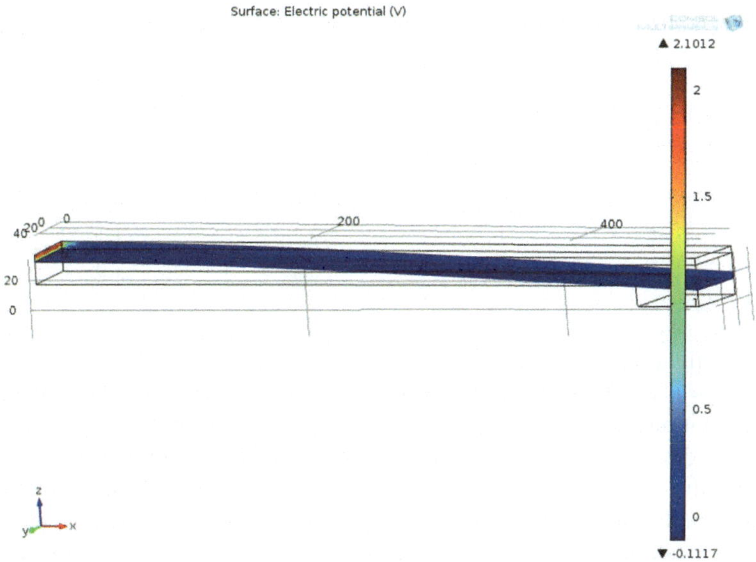

Fig. 7.119 Electric potential when $L_{canti} = 500\ \mu m$

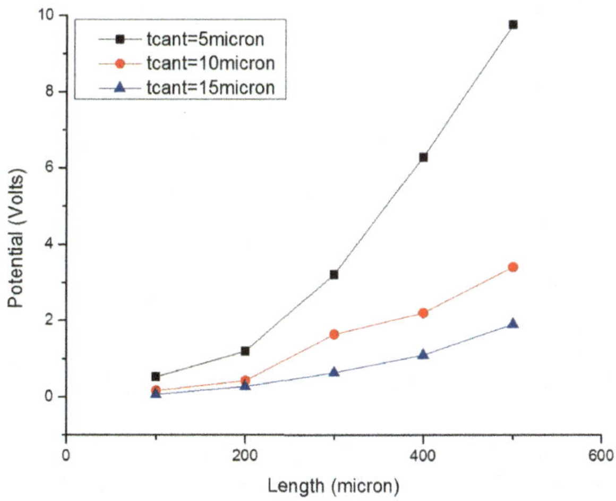

Fig. 7.120 Electric potential versus cantilever beam length

It can be seen from the above analysis that the length and thickness of the beam have an effect on the electric potential generated. From Table 7.7, it is observed that as the length of the beam increases from 100 to 500 μm, the electric potential generated by the cantilever increases from 0.52 to 9.75 V for a beam thickness of 5 μm. It can also be seen that as the beam thickness increases to 10 and 15 μm, the electric potential generated decreases. This is because as the thickness of the silicon beam increases it becomes stiff and reduces the displacement which reduces the maximum stress generated resulting in the lower electric potential generation.

Table 7.7 Potential versus cantilever length at different thickness

S. No.	L_{canti} (μm)	Electric potential (V)		
		$T_{canti} = 5$ μm	$T_{canti} = 10$ μm	$T_{canti} = 15$ μm
1	100	0.52	0.17	0.069
2	200	1.19	0.43	0.278
3	300	3.21	1.64	0.64
4	400	6.28	2.2	1.1
5	500	9.75	3.4	1.9

An investigation on the effect of structural parameters of the device on the output of the energy harvester has been done. The parametric study done shows the design and optimization part for an energy harvester device. Through this investigation, the range of the device operation can be investigated with different boundary conditions which give the nature of device operation. For any real-time application, the specific boundary conditions can be applied which will give the suitability of the device for that application. After the optimization of device parameters, the device can be realized using MEMS fabrication technology.

The manufacturer's authorised representative in the EU is Springer Nature Customer Service Centre GmbH, Europaplatz 3, 69115 Heidelberg, Germany. If you have any concerns regarding our products, please contact ProductSafety@springernature.com

Printed and bound by CPI Group (UK) Ltd, Croydon, CR0 4YY
30/09/2026
02225294-0003